By Earl B. Shields

Twentieth Edition - 1999

SHIELDS PUBLICATIONS
P.O. BOX 669 • EAGLE RIVER, WISCONSIN 54521

PRINTED IN U.S.A.

ISBN 0-914116-21-5

Greetings from "Possum Holler" . . . Most Peaceful Retreat in the Beautiful Arkansas Ozarks!

Earl B. Shields

RAISING *Earthworms* FOR PROFIT

By Earl B. Shields

A COMPLETE illustrated working manual for one of the most fascinating and profitable of home-money-making or full time business projects . . . one which requires no special skill or training . . . which may be launched with the smallest of investments and built to imposing proportions within the limits of a basement, garage or city lot . . . which any man, woman or partially handicapped person may conduct successfully as a part-or-full-time project.

Anyone who delights in outdoor activities, who likes fishing and fishermen, who loves the soil and has a yen for growing things . . . such a one will thrill to the fascination of this profitable part-time hobby or full-time business.

U. S. Department of Agriculture Photo

Showing relative sizes of common earthworms. Left to right, Native Nightcrawler, Field or Garden Worm, Diplocardia Earthworm, Green Worm, Manure Worm and Red Worm. The last two mentioned are those normally used in commercial production. The other four are not suitable for pit culture because of slow reproduction. Illustration, about 60% of natural size.

By Way of Introduction

MUCH water has flowed under the bridge since the first edition of this book was published. The earthworm industry has grown tremendously during the intervening years, with even larger prospects for future growth.

Despite the fact that hundreds of new growers enter the field each year, the skyrocketing demand for earthworms continues to outstrip production. There is still a severe shortage of really good worms for bait, breeding stock and for horticultural demands.

The earthworm industry has learned much in recent years about successful, economical earthworm production. New and better methods have been introduced. New types of worms have been developed, new ideas in housing, feeding and harvesting, new methods for reaching the ever-widening market.

Those are the reasons for this revised and enlarged new edition of "RAISING EARTHWORMS FOR PROFIT" which in previous editions has sold more than 400,000 copies and which we like to believe has had an important share in the rapid development of this multi-million dollar earthworm industry.

In this edition we have retained all of the basic materials of previous editions, but have added a great deal more that is new and important. Many new illustrations have been added also because we believe, with the Chinese, that "One picture is worth ten thousand words".

This book lays no claim to be a complete manual on earthworms and earthworm growing. It purposely omits, for example, the detailed scientific discussions of earthworm biology and history which may be found in other books. Its purpose is to cover all of the practical, down-to-earth factors in earthworm farming, plus valuable information on all phases of advertising and marketing . . . in short, HOW TO MAKE MONEY with earthworms.

It is the hope of the author that it will answer most of your questions about earthworm growing, and that you will find it helpful in launching or improving your own earthworm project.

Everyone is Fascinated by the Earthworm Story

THERE was a time, not too many years ago, when anyone who talked about raising earthworms was regarded as something of an "odd character" . . . someone whose mental machinery may have begun to creak a little.

His friends listened to him, loose-jawed, in utter amazement, incredulity, amusement or incomprehension . . . dependent upon their degree of familiarity with "the intestines of the earth", as the Greek philosopher, Aristotle, dubbed the earthworm.

But times have changed. Literally millions of people in the past few years have become familiar to some degree with the earthworm story, and have gained a vast respect for an industry that has grown so rapidly and so soundly that it has taken on something of the atmosphere of a "Horatio Alger" success story.

So we run a small classified ad in a few national magazines, offering "Free information on raising earthworms" and get 1,500 to 2,000 inquiries . . . from doctors, lawyers, bankers, clergymen, businessmen, farmers, factory workers, housewives, students, retirees, and a host of others in all walks of life.

An impressive number of such inquiries are from women, and "Never underestimate the power of a woman" in this amazing earthworm industry! Hundreds of successful earthworm hatcheries are owned and operated by women, with or without outside help. In hundreds of others, operated by husband and wife teams, the wife is the indispensable factor, acting as corresponding secretary, bookkeeper, purchasing agent and assistant picker and packer, sometimes as "guiding genius" of the project.

The launching of a new earthworm project, as a rule, follows a fairly consistent pattern. It is begun as a part-time or spare-time avocation or hobby, with the hope, perhaps, that it may eventually grow into a full-time living-income business . . . one that will allow the owner to cut loose from a weary time-clock or desk job to devote his entire time to a pleasant, healthful outdoor activity which has many rewards beyond mere income. For many of these beginning growers, DREAMS DO COME TRUE!

For example:

A Texas newspaper man, who operated an earthworm hatchery as a sideline for several years before cutting loose from the newspaper treadmill, became one of the largest and most successful growers in the Southwest.

The owner of an impressive chain of dry goods stores, bored with "standing behind a ribbon counter", started an earthworm hatchery a number of years ago, supervised it while hired help managed the stores, built his earthworm sales to imposing proportions, then sold the chain of stores. He operated one of the largest hatcheries in the United States, with sales of $40,000.00 to $50,000.00 per year.

An Oklahoma woman, without benefit of formal education and with slender resources, built an earthworm business from a modest start, supported her family, educated her children and built a new home . . . all out of earthworm earnings.

A Kentucky clergyman, faced with the limitations of a minister's income, established an earthworm hatchery to finance his son's college education, wrote an excellent earthworm manual based on his experiments and research . . . then, with the educational objective accomplished, sold the hatchery and the book rights.

The appealing thing about the earthworm venture, to those of limited means, is the fact that it can be started with a twenty dollar bill, plus a little labor, if one is willing to start slowly and let Nature take her course. If you are a bit impatient and want to get away to a fast start, a larger initial investment will put you in position to start selling in volume the following season.

Just one word of caution: don't expect your project to be all play and no work. No business venture can succeed without sound planning and diligent effort, and this one is no exception.

The work, however, is not too difficult. You don't have to be a genius and you can tailor your project pretty much to your own specifications as to investment, physical ability, and time available for it.

What is a "Domesticated" Earthworm?

DOMESTICATED earthworms, as the term is usually understood within the industry, are pit-bred worms, specially fed, watered and cared for to increase their natural size and their natural rate of reproduction. They are larger than the native red or manure worm, but retain its toughness, liveliness, attractive coloring and its prolific breeding capacity.

The red worm comprises the greater part of all commercially grown worms . . . possibly 80 to 90 percent at the present time. It is so much easier to raise than any other and the initial cost of getting it established is so much less that most growers are partial to it, while always seeking ways to grow it larger.

Red worms, in addition to being excellent bait worms, are hardy individuals, adaptable to a wide range of climatic and soil conditions. They need no housing other than a shed roof in most areas, though in extreme northern areas where the growing and breeding season is very short, it is advisable to raise them in heated buildings. They can stand any amount of cold except a solid freeze.

Earthworm buyers are sometimes confused, and understandably so, by the variety of names used by various breeders in describing their red worms . . . Red Hybrids, Red Wigglers, English Reds, Egyptian Reds, California Reds, Michigan Reds, Ozark Red Hybrids, Red Gold Hybrids, and a host of others.

These, for the most part, are individual "trade names" adopted by various breeders to designate their own particular stocks and to give them some distinguishing characteristic around which to build an advertising campaign . . . perfectly legitimate, of course, and an excellent sales policy.

The worms so described, however, are basically the same, derived from the same ancestral wild strains. Some domesticated worms, naturally, are better than others, but that is due in almost every instance to better feeding, handling and general good care.

There is one different type of red worm, with transverse yellow stripes at each segment, usually called the Red Wiggler in the South. The red worm is also called Red Wiggler by many breeders, but more often Red Hybrid in the northern and western states. The two types are equally good and similar in size, liveliness and breeding qualities. They are often found together.

A somewhat larger variety of domesticated worm is the Brown Nose Angleworm, light reddish-brown in color and about one-third larger than the red worm. It is thicker bodied, similar to a garden worm . . . a good breeder and shipper. Some breeders have done very well with it, but it seems not to have gained widespread popularity within the industry.

A worm that has gained in popularity in recent years is the grey nightcrawler. It, like other worms, has acquired different names such as "Hula Worm," "Swamp Wiggler," and "Jumper." According to Ellis Lake, a grower in Alabama, this worm is hardy, can be shipped easily and will stand both heat and cold. It is a plump, tough worm, sometimes growing to six inches in one year, and at an average weight of 3½ pounds per 1,000.

One large worm of the nightcrawler family is being raised successfully in pits outdoors in the deep south and indoors, under strictly controlled temperature conditions, in the northern states. It is variously called African Nightcrawler, Ethiopian Black, Florida Swamp Worm, and by other variations. It is harder and more costly to raise than the red worm, but commands premium prices on account of its size. You will find a chapter on the African Nightcrawler elsewhere in this book, with complete information on its raising and care.

HYBRIDS OR NO HYBRIDS... DOES IT MATTER?

Many conflicting views and theories have been expressed by writers and researchers as to whether there is, in fact, such a thing as a hybrid earthworm.

Dr. George Sheffield Oliver, in his book "Our Friend the Earthworm", recounts many years of experiment and research devoted to cross-breeding . . . an attempt to cross the orchard worm and the manure worm, or brandling, for the purpose of obtaining the better qualities of both. He maintains that, after five years of experiment, beset by many discouragements and set-backs, he did indeed create such a cross which he called the "Soilution" earthworm.

Whether or not such a hybrid exists is not very important. It may be just as notable an achievement if our fat and sassy domesticated earthworms of today are merely "educated" manure worms, developed through good feeding, good breeding and good management from the original little "stinkers" that infest manure and compost piles.

Just as wild razorback hogs have been developed through selective breeding into good meat hogs, or wild cattle into fine dairy animals, so the tough little wild worms may have been brought to their present state of development. It's a pretty sure bet that neither the fish nor the fishermen care a hoot about the ancestry of a fat, juicy bait worm!

A Good Way to Go Crazy!

UNLESS you are a mathematical genius, don't try to figure out the potential increase from your initial stock of mature breeders or pit-run units. The astronomical figures as you progress will give you a severe attack of "heebie jeebies".

To show you what we mean, we have prepared a very conservative table of figures indicating what each unit of 1,000 breeders, if allowed to increase undisturbed, might be expected to produce in the way of increase within a period of one or two years.

By "very conservative", we mean that we have credited each mature breeder with only TWO egg capsules per month, whereas, under favorable conditions, each worm should produce at least four capsules per month. And we have credited each capsule with hatching only TWO worms, whereas each fertile capsule normally hatches from two to twenty young worms, with an estimated average of at least four.

In other words, the following table shows about ONE QUARTER of the normal increase under ideal conditions, which we believe is conservative enough to offset most of the hazards of mortality losses, poor conditions, or just plain "bad luck".

Unfortunately, or perhaps we should say FORTUNATELY, in this instance, there is a fairly wide gap between theory and practice. The operator inadvertently makes mistakes in preparing bedding or in feeding and watering; cold winter temperatures slow down breeding activity; a hard freeze destroys the fertility of many egg capsules; rodents, mites or other insect pests may take their toll . . . all of these things may play a part in Nature's plan for balancing production.

So don't get the idea that you are going to be overrun by worms. You will not be operating under ideal conditions and you will also be selling breeder size worms which will reduce the actual number of worms and capsules produced.

A MIRACLE OF MULTIPLICATION

The following table is based on a single colony of 1,000 breeder earthworms (started in December) under controlled conditions and temperatures that permit continuous breeding throughout the year. It is based upon the premise that each worm lays only TWO capsules per month and that each capsule hatches only TWO worms.

Starting Stock (Dec.) 1,000 breeders

DATE	TOTAL BREEDERS	TOTAL WORMS (Including Breeders)	CAPSULES PER MONTH (Hatch in 21 days)
Jan.	1,000	1,000	2,000
Feb.	1,000	5,000	2,000
March	1,000	9,000	2,000
April	1,000	13,000	2,000
May	5,000	17,000	10,000
June	9,000	37,000	18,000
July	13,000	73,000	26,000
August	17,000	125,000	34,000
Sept.	37,000	193,000	74,000
Oct.	73,000	341,000	146,000
Nov.	125,000	633,000	250,000
Dec.	193,000	1,133,000	386,000

More than 1¼ million worms and egg capsules within one year from a start of just 1,000 breeders

DATE	TOTAL BREEDERS	TOTAL WORMS (Including Breeders)	CAPSULES PER MONTH (Hatch in 21 days)
Jan.	341,000	1,905,000	682,000
Feb.	633,000	3,269,000	1,266,000
March	1,133,000	5,801,000	2,266,000
April	1,905,000	10,333,000	3,810,000
May	3,269,000	17,953,000	6,538,000
June	5,801,000	31,029,000	11,602,000
July	10,333,000	54,233,000	20,666,000
Aug.	17,953,000	95,565,000	35,906,000
Sept.	31,029,000	167,377,000	62,058,000
Oct.	54,233,000	291,493,000	108,466,000
Nov.	95,565,000	508,425,000	191,130,000
Dec.	167,377,000	892,685,000	334,754,000

A GOOD WAY TO GO CRAZY!

A total of more than ONE BILLION* worms and egg capsules by the 24th month from a start of 1,000 breeders.

*We've carried the table this far as a "mathematical exercise" only. It is totally impractical as a working proposition because such a huge stock would require 10,000 to 15,000 average size pits, probably more than the world's total domestic earthworm population at this time!

MANY who read this book will be interested in the breeding of earthworms on a limited scale for personal use in soil building, or to provide a continuous supply of bait for angling.

Most readers, however, will be concerned with earthworms as a source of profit, either as a remunerative part-time hobby or as a full-time income project. It is for the latter, particularly, that this data has been prepared.

The market for earthworms is BIG and growing bigger so rapidly that even the hundreds of new breeders who enter the field each year are unable to meet the demand fully.

The consumer market falls into a variety of outlets, several or all of which the successful operator will probably cultivate in the promotion of his business. The most important of these (but not necessarily in the order named) are:

1 - EARTHWORMS FOR BREEDING STOCK

2 - EARTHWORMS FOR BAIT

3 - EARTHWORMS FOR SOIL BUILDING

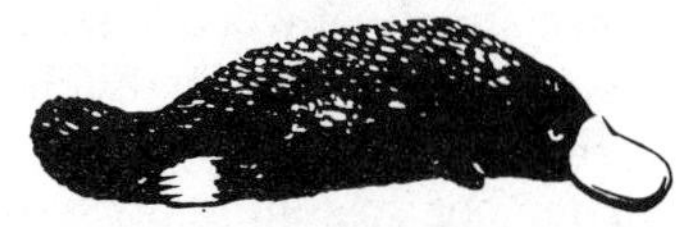

4 - EARTHWORMS FOR ZOOS, LABORATORIES, AQUARIUMS, GAME BREEDERS

YOUR MARKET FOR EARTHWORMS

The market is WIDE . . . nationwide for the average breeder . . . worldwide for those who are equipped to ship by air to foreign markets. Not many breeders, however, care to bother with foreign shipments, which are usually small, troublesome and not too profitable because of the amount of red tape involved.

The earthworm breeder is by no means confined to the limits of his local markets. Earthworms are good shippers and are easily packaged for parcel post or United Parcel transportation. United Parcel is best in most instances — cheaper, faster and usually more convenient than parcel post.

NO. 1 MARKET—EARTHWORMS FOR BAIT

This is a huge ready-made market that is easily reached with a minimum of effort and expense. We don't know how many fishermen there are in the United States, but we do know that they spend more than $200,000,000.00 a year for fishing licenses alone and several times that amount for live bait. It's a terrific market, and one that is steadily expanding with our growing population.

Unquestionably, America's favorite live bait (and we think, perhaps, the favorite of the fish also) is the dependable, fish-taking earthworm. Even the most fanatical devotee of plug or fly likes to take along a few earthworms, "just in case"!

You don't have to live in a good fishing or resort area in order to sell earthworms for bait. More bait worms are sold and delivered by mail than are sold to fishermen directly from the pits. Most bait shops, tackle stores and fishing camps get their supply of worms by mail, as do thousands of individual fishermen.

Many of the larger bait dealers contract well in advance for the season's supply of worms, some of them for as many as 25,000 to 50,000 per week for the season. Such contracts are usually made in January or February, or sooner.

Your heavy shipments will not begin, however, until sometime in May or June when the fishing season gets into full swing, or when fishermen begin to find it difficult to dig their own worm supply. (When the soil gets dry and hard the native worms go down to China!).

"NO LIFE my honest scholar — no life so happy and so pleasant as the life of a well-governed Angler; for when the lawyer is swallowed up with his business, and the statesman is preventing or contriving plots, then we sit on cowslip banks, hear the birds sing, and possess ourselves in as much quietness as these silent silver streams which we now see glide so quietly by us. Indeed, my good scholar, we may say of Angling as did Dr. Boteler of strawberries: 'Doubtless God could have made a better berry, but doubtless God never did.' And so, if I might be judge, God never did make a more calm, quiet, innocent recreation than Angling."

Izaak Walton in "The Compleat Angler"

YOUR MARKET FOR EARTHWORMS

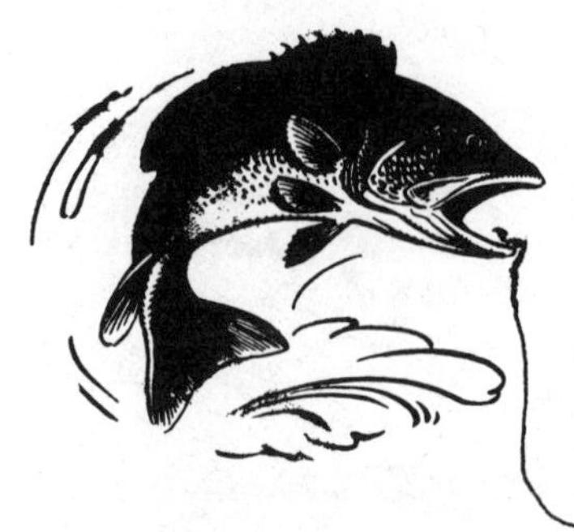

As time goes on you will build up a steady clientele of dealer customers who, if you ship only fine quality bait worms, will look to you for their supply from season to season. To get that clientele established you will probably do some classified advertising in the local and national outdoor sports publications. More about this in the chapter on advertising and sales promotion.

Retail prices of bait worms vary widely in various fishing areas, from as little as $2.00 per fifty to as much as $2.00 per dozen and more depending on kind of worm and size, the location, the lack of abundance of good bait worms and the fairness or greediness of the dealer. Somewhere midway between these extremes is probably a fair average

Wholesale prices will vary also, as they should, to correspond with prevailing retail prices. The dealer should expect a discount of approximately 40% off the retail prices if the worms are supplied in cups or cartons, ready to push across the counter . . . approximately 50% if supplied in bulk, where the dealer does his own packaging. Such discounts may be set up by a grower to fit local conditions if he is selling only in a local area.

The larger growers who ship nationally, however, sell at fixed wholesale prices without regard to retail prices. It is up to the retailer to fix his own prices. Such growers, who sell for the most part in large quantities from 10,000 up, will quote prices around $10.00 to $15.00 per thousand, postpaid anywhere in the United States. In many instances the local grower finds it advisable to quote similar prices in order to meet competition.

Most breeders do not accept orders from bait dealers for less than 1,000 worms on account of the inconvenience of packaging and mailing small shipments. For retail mail order buyers, quotations are usually for 100 and up.

The trend in packaging worms for bait dealers is more and more toward retail size packages rather than bulk, using attractively printed new cartons for 25, 50 or 100 worms, or whatever the trade demands. The old sloppy method of dispensing worms in old tin cans or salvaged containers of other kinds is on the way out. The live bait business has grown up over the past several years, and has assumed the dignity a buyer might expect to find in any other modern retail business.

It is good advertising, if volume justifies it, to have your cartons attractively printed at the factory, featuring your firm or brand name, with a suitable illustration and some printed instructions as to how to hook and use the worms most effectively.

If that practice is too costly for your volume, a printed label used on inexpensive plain cartons will serve almost as well. Such a label gives your hatchery an "identity" and suggests permanence and business stability. More about this in the chapter on Advertising and Selling.

ARKANSAS LUNKER BROWN TROUT—CAUGHT ON RED WORM

Fishing with spinning tackle, an eight pound test line and a red worm, Joe H. Todd of Ardmore, Oklahoma, caught this record smashing German Brown trout which weighed 9 pounds and 12 ounces. It measured 27 inches in length and 17 inches in girth. Todd was wading and casting when he hooked the trout on the North Fork River, a short distance below Lake Norfork Dam, near Mountain Home, Arkansas.

Photo and data, courtesy of The Baxter Bulletin, Mountain Home, Arkansas.

RED WORMS vs. OTHER BAITS FOR CATCHING FISH

FISHING is fun, no matter how, when or where you find it . . . no matter what kind of tackle, bait or lure you use. It doesn't matter whether you angle from boat, pier, or shore . . . from the grassy banks of a meadow brook or the rugged reaches of a wildly roaring mountain trout stream. It's even more fun if you catch fish, an accomplishment in which the novice sometimes confounds the experts.

You'll recall many a story or cartoon quip about the elaborately equipped fly rod sportsman, weary of limb and empty of creel at the end of a fruitless day, who comes upon a touseled urchin with a wild cherry sapling over one shoulder and an unbelievable string of beautiful trout over the other.

Maybe you've personally experienced a scene like that when you were a youngster, as have I. Did you ever poke a slender sprout through the bushes over a dark and promising pool and thrill to the excitement of a spine tingling tussle with hungry and hard-hitting speckled brook trout? Such are the experiences of fondest memory from my own boyhood.

Bait? Well, it was usually whatever the river bank afforded . . . earthworms rooted out with a crooked stick, grasshoppers and crickets from nearby fields, grubs from rotting logs, crawfish and hellgrammites from beneath overturned stones . . . any live bait that could be impaled on a hook. The trout loved them all!

When it comes to lures, the addict fly or plug fisherman may disdain the use of live bait as being beneath his skill. And even the minnow fisherman may feel that some sort of stigma attaches to the use of earthworms for bait.

Not so the most famous angler of modern history, Izaak Walton, who records in his classic "The Compleat Angler":

". . . and that one day especially, having angled a good part of the day with a minnow, and that in as hopeful a day, and as fit a water as could be wished for that purpose, without raising any one fish; I at last fell to it with the worm, and with that took fourteen in a very short space."

Good old Ike Walton! Just a couple of days ago, as this is written, this writer got his limit of Rainbows, wading and casting on the White River, with red worms . . . all in 1½ hours.

It's a pretty safe guess that fish haven't changed much in their feeding habits or their food preferences since the days of Izaak Walton, and today, as then, I have no doubt that more fish are taken with earthworms each season than with any other single type of bait or lure.

There may be nothing subtle about luring a trout, bass or crappie with a lively, wriggling red earthworm bait, but the fight is just as exciting once you get him on, especially if you're using an ultra light rod with a small hook and a four to six pound test line.

That's the way to fish with earthworms for maximum results and sport, a trick that some fishermen have not yet discovered.

Down in our Ozark rivers and lakes, where lunker bass run as heavy as 8 to 12 pounds, many fishermen use enormous hooks, apparently in the conviction that large hooks should be employed in angling for large fish.

Obviously you can't bait a "hay-hook" with a comparatively small earthworm, however fat and juicy a morsel he might seem to a hungry fish. Nor can you wad several small worms, or a large nightcrawler, onto such a hook and make it look like anything a fish ever dreamed about as a tempting supper dish. A dead worm is a dead bait.

The small hook, properly baited and tied to light line, is the answer. It will frequently take the big fish more readily than a larger bait, and you'll land them too, if you have the skill and know-how. That's where the sport comes in. Anyone can "horse in" a big fish with a heavy line, a wire leader and a gang of hooks on a wood plug, but it takes a real fisherman to do the job with light tackle!

In baiting your hook with a red worm, don't thread him on like a joint of macaroni . . . hook him lightly under the clitellum (the thick band near his head), and then again directly into the head of the worm. He'll live and wriggle for half an hour or more, if the fish leave him alone that long. If they do that, however, you may feel reasonably sure you're fishing in the wrong spot.

Earthworms, of course, are not always the best bait. Sometimes the big fish are hitting minnows; sometimes flies, poppers or plugs. But it's a wise precaution, no matter what type of lure you favor, to have along a supply of good, dependable red worms, "just in case".

> "I envy not him that eats better meat than I do, nor him that is richer, or that wears better clothes than I do: I envy nobody but him, and him only, that catches more fish than I do."
>
> Izaak Walton in "The Compleat Angler".

DID YOU EVER SEE TEN MILLION WORMS ALL AT ONE TIME?

WE ASKED a leading bait processing company for permission to reproduce this fascinating photograph of a "beautiful gob of worms." We thought you'd like to see it. This company does not grow worms, but buys them for packaging in plastic or glass from growers who supply the worms already preserved in a special formula. The packaged worms, along with other preserved baits, are sold through sporting goods stores. While this does not constitute a large worm market, it is one that should be considered. Please do not ask for the name of the above company. They have adequate suppliers.

Most fishermen prefer live bait. Preserved bait, however, will look a lot like live bait and by treating it with a special odor solution, it can attract fish almost as well as live bait.

While some growers deal through companies that specialize in the sale of preserved baits, other growers process and market their own. It may be the answer to the question "What to do with a surplus?" at the end of the season. For those baits that are in short supply certain times of the year, this method makes it possible to store baits when they are plentiful for those times when they are scarce and higher priced.

A small instruction manual titled "How To Preserve Bait of All Kinds", written by Charlie Morgan, tells how to preserve worms, frogs, crickets, crawfish, etc., and describes packaging and shipping containers. The booklet is available for $2.00 from Shields Publications, P.O. Box 669, Eagle River, WI 54521.

EARTHWORMS FOR BREEDING STOCK

BREEDING stock may well represent the No. 2 market for earthworms, though we have no conclusive figures to prove it. We do know that many large growers specialize in breeding stock and that it represents the greater part of their sales volume; also that most other breeders sell a considerable number of breeding stock orders from time to time.

There are two primary sources of demand for breeding stock:

1 - There are hundreds of beginning breeders who start new hatcheries each year; some in a small way and some large. Recently, the trend has been to sell in units of worm beds containing approximately 100,000 pit-run worms. The sale generally includes the bed framework, bedding, initial feed, instructions, and consultation assistance if help is needed.

Many beginning growers, however, are cautious and will start experimentally with an order of 5,000 or 10,000, possibly less. These small starters, if their worms do well, will quickly come back for more stock, usually to the same grower who first supplied them. The total for all of these new breeders is large.

The grower who advertises and sells breeding stock to beginning breeders usually enjoys a slightly better profit than if he sold to bait shops, for example. Prices are less competitive for hand-selected banded breeders or pit-run, and as a rule they can be delivered in gallon containers, which represents a considerable saving in packaging cost.

2 - The second source, and a large one, consists of established growers whose sales deplete their stocks to the point where they have to go into the open market for large quantities of worms for restocking their pits, or for supplying their customers. In some instances, such a grower will buy a million or more worms from other growers in the course of a year, usually pit-run stock at low bulk-wholesale prices. Such sales do not carry as large a margin of profit as do sales direct to retail outlets, but they are a lifesaver for the grower who finds himself with a surplus that he wants to turn into cash.

There are a few growers who specialize in supplying other growers, in preference to bucking the general market. Even at the low bulk prices at which they sell, they make a satisfactory profit because they completely eliminate advertising, selling and packaging costs. Frequently a buyer brings his own truck and buys a truck load or the contents of one or more entire pits at one time. Some growers, in fact, feature whole pit sales.

EARTHWORMS . . . MASTER SOIL BUILDERS

U.S. Department of Agriculture Photo

COMMERCIAL earthworm growing owes its origin to the dedicated missionary work of pioneer scientists and researchers whose consuming interest was the value of earthworms in soil improvement . . . such men as Charles Darwin, Sir Albert Howard, George Sheffield Oliver; and in more recent years, Dr. Thomas J. Barrett, J. I. Rodale and soil conservationist Henry Hopp of the U.S. Department of Agriculture.

We do not propose, in this manual, to enter into a scientific discussion of the importance of earthworms in soil building, or the methods employed in their use. For a complete discussion of this subject we recommend to you, for supplementary study, "Harnessing the Earthworm" by Dr. Thomas J. Barrett and "Let an Earthworm be Your Garbage Man", which includes a treatise by Henry Hopp of the Department of Agriculture.

Most scientists, zoologists, soil culturists and agricultural experts are agreed upon the importance of building up the earthworm population in the soil for maximum plant growth and production, and as a soil conservation measure.

Earthworms, as they burrow and feed, swallow great quantities of soil, digest it, extract its food value and expel the residue as worm castings, which are infinitely richer in nitrogen, phosphates, calcium and magnesium than the finest of top soil. All elements are water-soluble and immediately available as plant foods.

Equally important, earthworms constantly till the soil around root areas by their tireless burrowing. The burrows form the channels through which the root growth may reach down into the subsoil for minerals and moisture, and these channels also absorb rainfall quickly for storage in the soil instead of allowing it to run off, carrying away valuable topsoil with it.

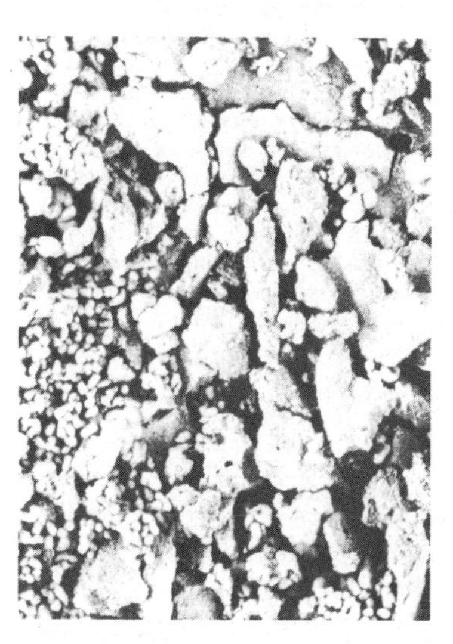

U.S. DEPARTMENT OF AGRICULTURE PHOTOGRAPHS SHOWING THE EFFECT OF EARTHWORMS ON THE SOIL

LEFT: Soil heavy, packed, almost impervious to water. Cakes under hot sun; hard to cultivate; restricts plant and root growth. No earthworms.

RIGHT: With earthworms; soil is granulated, light, crumbly; absorbs water; is easy to cultivate; gives plant life a chance to grow normally.

The "Wonder Worm Farmers" of Columbus, Ohio (avid compost-plus-earthworms gardeners) write of "12-inch Kentucky Wonder green beans, 10-foot tall tomato plants, 14-foot pole limas and 90 quarts of strawberries from a patch calculated to yield 30."

Dr. Thomas J. Barrett, famous earthworm champion, scientist, experimentalist and writer, tells of "carrots weighing 6 pounds each, parsnips weighing 4 pounds each, turnips weighing 7¼ pounds each, Irish potatoes at the rate of 1,200 bushels per acre, onions at the rate of more than 800 bushels per acre" . . . all of these on his Earthmaster Farms project, with the aid of his own earthworm livestock.

The late Dr. Thos. J. Barrett, much honored dean of the early earthworm writers and research scientists, here displaying some of the giant vegetables grown on his Earthmaster Farms in California . . . a four pound parsnip and a six pound carrot, typical of a wide range of fabulous results obtained in Dr. Barrett's experimental earthworm gardening . . . results which brought to Dr. Barrett's home a long guest list of distinguished official visitors from every part of the world.

The much quoted aphorism, "Tell me what you eat and I will tell you what you are", might be applied as well to flowers, shrubs, trees, fruits and vegetables, because plants and their edible products reflect in growth, flavor and nutritional values, the elements absorbed by their root systems from the soil in which they grow.

Foodstuffs grown in fertile soil, with plenty of organic matter, quickly composted by earthworm action and vastly enriched by earthworm castings, are bound to become an adventure in good eating and healthful body building.

Darwin, in "The Formation of Vegetable Mold Through the Action of Earthworms", says:

"In many parts of England a weight of more than ten tons of dry earth annually passes through their bodies and is brought to the surface on each acre of land; so that the whole superficial bed of vegetable mold (top soil) passes through their bodies in the course of every few years."

Sir Albert Howard estimates the annual weight of earthworm castings in well populated soil at as much as 25 tons or more per acre. He writes:

"Earthworms condition the food materials needed by the roots of plants. This is accomplished by means of their casts which in a garden soil in good condition may exceed 25 tons to the acre in a single year.

"The casts are manufactured in the alimentary canal of the earthworm from dead vegetable matter and particles of soil. In this passage the food of these creatures is neutralized by constant additions of carbonate of lime from the three pairs of calciferous glands near the gizzard, where it is finely ground prior to digestion. The casts which are left contain everything the crop needs . . . nitrates, phosphates and potash in abundance, and also in just the condition in which the plants can make use of them.

"Recent investigations in the United States show that the fresh casts of earthworms are five times richer in available nitrogen, seven times richer in available phosphates and eleven times richer in available potash than the upper six inches of soil. The earthworm is, therefore, the gardener's manure factory."

Dr. Barrett writes, "In the chemical and mechanical laboratory of the earthworm's intestine are combined all the processes of topsoil-building. The earthworm swallows great quantities of mineral earth, with all that it contains of vegetable and animal remains, bacteria and the minute and microscopic life of soil. In the powerful muscular mill of his gizzard, using grains of sand for millstones, the ingested material is thoroughly ground and mixed, as the abundant digestive secretions are poured in to exert their solvent and neutralizing action. Slowly the semi-liquid mass moves through the intestine, undergoing further mixing as it takes on valuable animal hormones and substances. Finally, it is ejected in and on the surface of the earth as castings; earthworm manure, humus, a crumbly, finely-conditioned topsoil, richly endowed with all the elements of plant nutrition in water-soluble form."

In George Sheffield Oliver's book, "Our Friend the Earthworm", the author sounds this note of precautionary advice: "The tyro earthworm farmer must realize that it takes time for plants, shrubs and vegetables to show the benefits derived from the persistent and efficient work of the earthworms around the root zones.

"Many forms of plant and vegetable life show a marked improvement in from 30 to 60 days after the earthworms have been placed in the soil around their roots; in some instances, however, an entire growing season is required to prove the full merit of this type of culture."

The foregoing text in this chapter pertains, in large degree, to those worms which are NATIVE to the soils in which they are found; garden and field worms whose population may be built up by increasing their food supply, by adding humus, by plowing under manure, stubble or other organic materials. The same reasoning may be applied to domesticated red worms, but with certain reservations.

THE ROLE OF DOMESTICATED RED WORMS IN SOIL IMPROVEMENT

U.S.D.A. photograph of two experimental barrels of identical soil — right barrel with earthworms and left barrel without. The spectacular difference in growth is due solely to the work of the earthworms in conditioning the soil and composting the organic matter within the soil to make it readily available to the plant roots. From Hopp and Slater, report in Soil Science.

The difference between "native" worms and pit-raised worms, for use in the garden and around shrubs and trees, lies in the fact that the red, or pit-raised, worms must be continuously supplied with organic food materials, animal or vegetable, and must be supplied with at least a minimum of moisture under dry or drought conditions, whereas the native worm is better equipped to shift for himself.

Native worms go deep in dry weather for food and moisture. The pit-raised worm, on the other hand, is a surface feeder whose natural habitat is a manure or compost pile. It cannot survive in the soil under adverse conditions unless food and moisture are supplied. If these elements ARE supplied, the red worm will live, multiply and perform his soil-improvement function as well as the native worm, or better; without them he will be valuable only as fertilizer.

Unless food and moisture CAN be supplied by the grower when Nature fails to do so, then the proper function of the red worm in soil improvement lies in the use of the well-worked and casting-rich compost from the worm pits, quite as important and effective as the use of the worms themselves.

In the pits, after a few months of habitation, the red worms will turn tons of compost into the finest potting or mulching soil that has ever been discovered,soil that will make your own potted plants or your shrubs and garden the envy of your community. Or you may turn it into cash by bagging and selling it, either direct or through florists, seed stores or garden supply stores. More about that in the chapter on "Supplementary Income".

EARTHWORMS...CAUSE OR EFFECT OF GOOD SOIL?

THERE'S something of a difference of opinion between two schools of thought as to whether (1) soil is good because it contains an adequate earthworm population, or (2) earthworm populations are adequate because the soil is good.

It is in fact an academic question because both contentions are correct; the whole matter sums up to a happy combination of the two.

Earthworms thrive and become abundant only in soil that is rich in organic matter on which they feed.

The earthworms, on the other hand, help to break down and "compost" the organic matter in the soil, bring up minerals from the subsoil, digest the soil that passes through their bodies and expel it in the form of casts that are tremendously rich in plant-food elements. They turn organic matter into water-soluble humus immediately available as plant food.

They aerate the soil with their burrows, which also form channels for root expansion and enable the soil to absorb and retain much moisture that would otherwise run off, carrying valuable topsoil with it.

Nobody cares very much which was first, the hen or the egg, inasmuch as each is dependent upon the other.

Good soil and earthworms, too, are interdependent. Good soil is definitely better for the presence of a large earthworm population, plant growth is sturdier and healthier, blossoms are more abundant and colorful, fruits and vegetables are better in flavor and more nutritious.

When the soil around trees, bushes, plants, or in the garden or other growing areas is impregnated with earthworms or earthworm spawn, it should also be fertilized regularly with organic matter such as manure, leaves, straw, kitchen or animal waste, so that the earthworms may thrive, multiply and perform their own important special functions in soil building and the promotion of plant growth.

SECONDARY OUTLETS FOR EARTHWORMS

THERE are various miscellaneous markets for earthworms with which the average grower will rarely come in contact but which, nevertheless, may prove to be profitable outlets for you if they happen to be available to you.

Most of these, because requirements are large, would find it too costly to buy all of their needs from commercial growers at retail, or even wholesale prices. They should be able to grow their own worms profitably, however, and so become possible prospects for breeding stock.

POULTRYMEN, for example, can greatly improve the diet and health of their hens, increase egg production and lessen their cleaning and "odor" problems by maintaining earthworm pits beneath the roosts or elsewhere in conjunction with their buildings, also using the pits for sprouting grains to give the hens access to the tender green sprouts.

GAME BIRD BREEDERS have much the same problems and requirements as the poultryman and can use earthworms in the same manner.

FISH HATCHERIES find earthworms a natural food for growing fish, and could well afford to raise earthworms in connection with their hatcheries. The byproducts of meat packing plants and fish canneries, formerly available at low cost to fish hatcheries, have pyramided in cost, due to their use in large quantities by the manufacturers of dog and cat foods. In fact, their cost has become almost prohibitive for the private operator. Possibly earthworms can provide the answer.

ZOOS require thousands of earthworms for various animals. AQUARIUMS and FROG FARMERS are also good prospects for earthworms for breeding stock.

LABORATORIES: high school, college and laboratory supply houses are users and processors of earthworms for laboratory experimental projects and class room work.

RABBITRIES have become BIG users of earthworms under their hutches as a solution to their cleaning problems, and also as a source of supplementary income. So large is this market that we have devoted a special chapter to it in the pages immediately following.

VERMICOMPOSTING is a new organic waste management market employing red worms in the reduction of household and commercial waste. It is perhaps the fastest growing new international market. Refer to the books "Recycle With Earthworms", "Let An Earthworm Be Your Garbage Man", "Harnessing The Earthworm", and the video "The Red Wiggler Connection" for good information on this subject.

Earthworms and Rabbits . . .

A NEAR-PERFECT COMBINATION

RABBIT breeders have discovered a vein of pure gold beneath their hutches where formerly they had only an unsightly and unsanitary accumulation of rabbit droppings, fly-breeding urine, unsavory odors and a lot of hard work involved in an attempt to keep the area clean and odorless.

Today, in hundreds of modern rabbitries, rabbit droppings are feeding millions of fat and sassy domesticated earthworms; the cleaning problem is all but forgotten, rabbitry-bred flies and objectionable odors have disappeared, and best of all, the breeder has a priceless new source of income at practically no increase in operating expense.

Rabbit manure, plus the wasted feed from the hutches, is one of the finest of all earthworm feeds. By building pits or bins beneath the hutches, and stocking them with pit-run earthworms, the worms consume and compost the droppings as they fall, turning them into a finely pulverized, odorless humus. It is the finest of potting or mulching soil, readily salable to gardeners and flower growers or through such outlets as nurseries, seed stores, florist shops or garden supply stores. This humus, in turn, absorbs the urine and deodorizes it, prevents moisture accumulations and discourages the breeding of flies and other insect pests.

Photographs courtesy of Avery R. Jenkins, Sunnyside Rabbitry and Worm Ranch

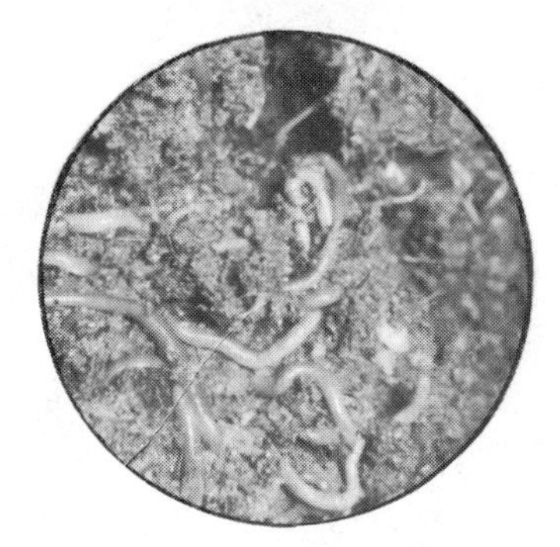

Bins or pits of cement, cinder blocks or 2 x 12 lumber, with provision for drainage, are built on the floor (if cement) or a few inches below the level of an earthen floor. They should be a few inches larger than the hutch area, to catch all droppings, urine and feed wastes. A few inches of compost, which may be a 50-50 mixture of rabbit droppings and peat moss, in the bottom of each bin, is sufficient for starting the worms. After that the rabbits will supply new food materials daily and the worms will do their own composting.

Breeding stock may be obtained from any earthworm grower, including many rabbit breeders who now raise earthworms beneath their hutches and have a surplus to sell to other breeders or to bait shops.

For the rabbitry, we recommend pit-run worms for foundation stock rather than mature breeders. In pit-run you get more worms for your investment, and they adapt themselves more readily than do mature worms to new environments and feeding.

The only work involved in maintaining the worms is to keep the pits moist (as in any other earthworm operation) and an occasional leveling off of the accumulated droppings. Heavy urine concentration hot spots should be removed as you find them. The contents of each bin should be forked over every two or three weeks to keep it loose, and to pulverize any encrusted manure deposits.

If eventually the bins get too full or too crowded with worms, the worms should be marketed to bait dealers, to other rabbitries for breeding stock, or to any other of the many outlets for earthworms. The surplus compost may be bagged and sold as previously indicated. No richer plant food can be found anywhere, and there is a lively demand for it.

From the standpoints of improved conditions, time and labor savings, and dual-income possibilities, the Rabbit-Earthworm combination is a NATURAL!

Photograph courtesy of Louis J. Cassata, Big "D" Rabbitry and Worm Hatchery. The pits and hutches are shaded by the fast growing Italian Squash vines.

WHICH IS THE BY-PRODUCT?

Under the above heading, Robert F. Thomas, North Carolina owner and operator of the famous Tarheel Rabitry and Worm Hatchery, contributed an interesting Rabbit-Earthworm story to Small Stock Magazine. By permission of the author, excerpts from that story follow:

"Much has been written in past years about small breeders who operated rabbitries of perhaps 25 holes, more or less. Many of us had to do backyard butchering and sell to neighbors. We threw the pelts away and gave what droppings we couldn't use ourselves to someone to get them off the lot. The argument was always coming up — were we, the small breeders, commercial growers or fanciers?

"Just because the rabbit breeder enjoys trying to develop an improved strain of rabbits, and is interested, also, in showing his rabbits, does not mean it is strictly a hobby, and that he does not care whether or not he shows a profit at the end of the year.

"The writer was one of those having to use the red ink, so after having read everything that had been published about combining rabbits and the domestic earthworm (and there was very little at that time), we decided to do something about it.

"We ordered 10,000 pit-run hybrid earthworms and dumped them under two double hutches (four holes), using 2 x 12 boards for frames. By late May we had so many worms in the bed that it was necessary to start moving them to other beds. By the end of the year there was 500 square feet well stocked with earthworms, and we had sold several hundred dollars' worth to fishermen and other rabbit breeders in our immediate area.

"Instead of trying to get someone to haul off our rabbit manure, we

Photograph courtesy of O.W. Van Camp, Ocean Breeze Rabbitry. Earthworms are kept in dropping pits which are open and easy to work.

Photo courtesy of Howard "Lucky" Mays, Happy Chick Animal Farm. Note arrangement of hutches for easy access to worm beds.

were trying to buy from other breeders. At this date every member of our club, and almost every breeder that I know in North Carolina, have included earthworms in their rabbitries, and I have to supplement my rabbit manure with other kinds of manure, mostly bought from dairymen and farmers.

"The Tarheel Rabbitry has answered more than a thousand letters about earthworms and rabbits, mostly from other rabbit breeders. We are proud of the fact that although we had to burn some midnight oil, we tried to answer everybody as honestly as we knew how.

"The time has arrived when all rabbit breeders who wish to take full advantage of the profits to be made in the rabbit business, and who have not introduced domestic earthworms under their hutches, are just not keeping up with progress, in our opinion.

After applying ourselves diligently to our worms and our rabbits, my wife and I sometimes wonder which is the by-product . . . the rabbits or the earthworms!"

Another successful rabbit-worm raiser is Howard "Lucky" Mays, of Mississippi. His book "Raising Fishworms With Rabbits" is the only book on this subject to our knowledge. It tells of his experiences in establishing a rabbit-worm farm with step-by-step advice to the person who wants to raise earthworms under rabbit hutches. It is easy to see why many such farmers find that the earthworms make the difference between a loss and a good profit.

Planning Your Earthworm Project

AS IN any other undertaking, however large or small, careful advance planning of your earthworm project will pay off; in saving of time, money, labor, in avoiding costly errors. Such planning will give you the confidence and assurance that will get you away to a successful start. May we suggest the following preliminary steps:

1 - Read everything worthwhile you can find on the subject; good books by reliable authors. They will repay their costs many times over . . . by guiding you in all of the important details of housing, pit construction, breeding, feeding, beddings, packing and packaging, advertising and selling . . . everything you'll need to know to raise earthworms successfully, avoid pitfalls and minimize risks.

2 - Check your resources for growing earthworms by asking yourself a few questions: Will I like this kind of work? Am I physically able to handle it? Can I spare the small investment required, and can I afford to hold on for a year, or thereabouts, until a return may be expected? Can I expect a good local market for my worms, or if not, can I afford to make a small investment in classified advertising and literature to win a share of the national mail order market? Are my facilities suitable . . . basement, back yard, or larger space for expansion?

3 - If possible, visit two or three going earthworm hatcheries near you, or farther away if necessary, to observe earthworm growing in practice. Most growers are glad to show you their hatcheries, tell you about their methods and answer your questions. You may find one from whom you'd like to buy your own foundation stock. (If you find one who WON'T answer questions, who infers that his methods are exclusive, deeply dark and mysterious, forget him . . . he is not typical of the industry.)

4 - Purchase a copy of the latest edition of the "Earthworm Buyer's Guide" (see book section) which lists leading earthworm breeders, and write to a few of them for literature and prices. Many of them carry display ads in the back of the Buyer's Guide with details of their offerings and sometimes prices.

5 - Lay out your project on paper first. Figure your costs carefully. Then, so you may begin selling in volume the following season, begin with as many breeders as you can handle or afford.

How to Get Started With Earthworms

ONE of the most appealing things about starting an earthworm project is the fact that you can begin it on any scale that suits you best. You can tailor it to fit your purse, your physical resources, your available free time . . . and your temperament.

You may regard it as a "hobby" if you choose, as a part-time extra-income plan, as a retirement project, or as the beginning of a new full-time living-income vocation that will get you out-of-doors, and eventually enable you to get away from the time clock grind or the handicaps of a desk job.

You can start on a twenty dollar bill, if you are willing to start slowly and "let Nature take her course". Or if you're in a hurry, you may invest anywhere from a few hundred to a few thousand dollars and be in business RIGHT NOW. Many a large hatchery, however, grew out of a modest start. In most instances, it is the better part of wisdom to be cautious; to start on a scale you know you can maintain; to get the "feel" of the business before you invest too heavily.

Possibly somewhere in between these extremes would represent the most sensible course, a start large enough so you won't "get the jitters" waiting for your business to reach the stage of cash returns, but small enough to hold down your initial investment, and minimize any possible risk.

Having decided to make a start and having planned your project, the first step is to prepare your facilities for the worms, before you purchase your foundation stock; a few boxes or a pit or two, compost for filling them, the few necessary tools, all of which should be ready before your breeding stock arrives. The preparation of compost or bedding for your pits or boxes is covered in another chapter.

In the meantime you will have contacted one or several good sources for breeding stock. You may find among these breeders a rather wide discrepancy in prices quoted, but price alone is not always the best guide to value. Most breeders are reliable and will tell you in their literature exactly what kind of stock to expect. If in doubt, order a sample thousand worms in advance of your main order.

You can start with mature banded breeders, which will get you off to a faster start, or you may order pit-run (mixed sizes, or all small worms — make sure which) at considerably lower cost per thousand. Pit-run, if it contains a considerable percentage of large worms, plus some egg capsules, may in fact be an advantage (1) because the young worms will adapt themselves more readily to their new environment, and (2) because you'll probably get more worms than you pay for. Nobody actually tries to count every tiny worm in such a shipment and is apt to give you the benefit of a liberal guess. Pit-run will take a little longer, possibly 30 to 60 days, to reach the stage at which you would be starting with banded breeders.

When your breeding stock arrives, the worms will probably be "balled up" in a solid mass within each container. Dump them out on top of the prepared bedding, gently disentangle and spread them over the surface. They'll go down quickly and within a couple of minutes will disappear completely. For a few days they may be inactive, but will soon begin feeding and breeding, if conditions and temperatures are favorable.

A WORD OF CAUTION:

Don't overlook the fact, in planning your project and estimating your capital outlay and your operating costs, that earthworms have to be SOLD as well as reared, and that some selling expense will be involved. For more on this please refer to the chapter entitled Earthworm Advertising and Selling.

HOW TO SPEED UP YOUR EARTHWORM PROGRAM

If you make a conservative start with earthworms, as many do, the demand for worms may quickly outstrip your production. In that event, you may find it advisable to buy and resell bulk wholesale worms from some other breeder who makes a specialty of supplying such worms to other breeders . . . usually mixed sizes, at costs low enough to enable you to make a satisfactory profit on them.

From such shipments you can sort out and sell the bait or breeder size worms, dumping the smaller worms into your own pits to grow. In this way, you can take care of all demands and build up your customer list while you are waiting for your own pits to get into volume production. It's a good way to get off to a fast and profitable start, and to put your earthworm project on a paying basis right from the beginning.

STARTING WITH PROPAGATION BOXES

THE only use for which small boxes like these are recommended is that of a small home or experimental project, for the grower's personal use, or as a means of "getting acquainted" with the breeding, feeding and living habits of earthworms as a preliminary to establishing full scale production in pits.

As a serious start for volume production, we recommend starting instead with at least one pit, indoors or outdoors. You will find it a lot less work to handle, feed and work one pit than a dozen or so boxes,

Illustration shows the recommended type of propagation boxes and method of stacking. The use of dividers between boxes (shown in circle) is optional. The same separation may be achieved by nailing cleats on the bottom of each propagation box. The base for stacking the boxes is made of 2 x 6 inch lumber and may be any length required for the number of boxes to be stacked.

Courtesy Earthmaster Farms

which multiply rapidly and soon become a problem to handle and store.

Many beginning breeders do like to start in this way, however, and it has certain advantages. Egg capsule production reaches its highest peak under the high concentration of breeders possible in boxes, which doubtless produce more worms per cubic foot of compost than any other method. Also, boxes offer an opportunity to experiment with various lots of breeders on a small scale, to correct errors in feeding and handling, without danger of substantial loss involving the entire stock.

Ready-made lug boxes are not available so when making them keep in mind that they should be small enough for easy handling. They should also be uniform in size so they may be stacked for economy of space. They should be tightly constructed, but with small holes or narrow cracks in the bottom for drainage. It is advisable to provide a false bottom of lath to reinforce the bottom of the box and make the damp compost come away clean when the box is dumped. A small cleat on each end of the box, for finger grips, will make it easier to handle.

Boxes should be set or stacked off the floor to prevent worms from working their way out through the drain holes or cracks.

WORKING WITH PROPAGATION BOXES

COMPOST, or bedding, for propagation boxes may be the same as that described for earthworm pits in another chapter. In the bottom of each breeder box, spread a layer or two of burlap (old gunny sack) or a one inch layer of alfalfa, straw or dry leaves to aid drainage and provide supplementary worm food. Newspapers can also be used for this purpose. Fill the box ⅔ full of damp bedding, and on top of that dump about 500 breeders or 600 to 700 pit-run worms; worms in all stages of growth. In a few minutes they will burrow down out of sight. Spread a thin layer of poultry mash on top of the bedding and water lightly. Then cover the surface with wet burlap to keep the contents of the box from drying out.

It needs no further care, except for an occasional feeding and sprinkling through the burlap covering, for a period of 21 to 30 days, when the contents of the box (if stocked with mature banded breeders) should be ready to dump and divide. Sprinkle supplementary food on top of the compost as often as the worms consume it. To keep the worms from crawling out, lights should be kept on over the boxes.

Examine each box at the end of 21 to 30 days to determine the number of egg capsules and young worms. If they are plentiful it is time to dump and divide the box.

With the contents of the box on the top of a smooth table, rake it with the fingers into a cone-shaped pile and let it stand for a while in bright light so the mature worms will work down to the bottom of the pile. Then, little by little, transfer the compost with its egg capsules and baby worms back into the original box, removing any mature worms that may be encountered. Have a new box ready with fresh compost; put the mature breeders into it, add food, cover with burlap and sprinkle, as in the first operation. That box should be ready to dump and divide again in another 21 to 30 days.

The box of spawn (capsules and young worms), with additional compost added if necessary, may remain undisturbed, except for an occasional feeding and watering, for about 90 days. By this time it should be swarming with young worms hatched from the capsules, plus some that have developed to breeder size, indicated by the band or clitellum. If the box seems to be crowded, dump and divide it, again transferring the compost and small worms back into the original box and preparing a new box for the breeder size worms, which will then take their place in the 21-30 day rotation.

YOU DON'T NEED A "FANCY" OR COSTLY SET-UP TO START YOUR WORM PROJECT

A FEW wash tubs, an old stock tank cut into segments, oil barrels cut in half, a wood pit built around a tree, a board pit, supported by stakes in a dirt floor shed . . . any of these, or any number of substitutes you may dream up for yourself, can get you off to an inexpensive start.

A wash tub, for example, can be stocked with a thousand breeders and will accommodate as many as 3,000 mature worms before you have to divide them.

You can build your large outdoor pits, if you wish, while the worms are multiplying in these low-cost beds, and by the time the larger pits are ready and filled with bedding, you'll have an abundance of good worms for stocking them.

Many an impressive and highly profitable big-volume worm hatchery has grown from such a beginning, paying its own way as it went along.

The photographs on this page were borrowed, with permission, from the literature of Oakhaven Farms, a Texas hatchery that began small and grew large, by virtue of good worms and good selling methods.

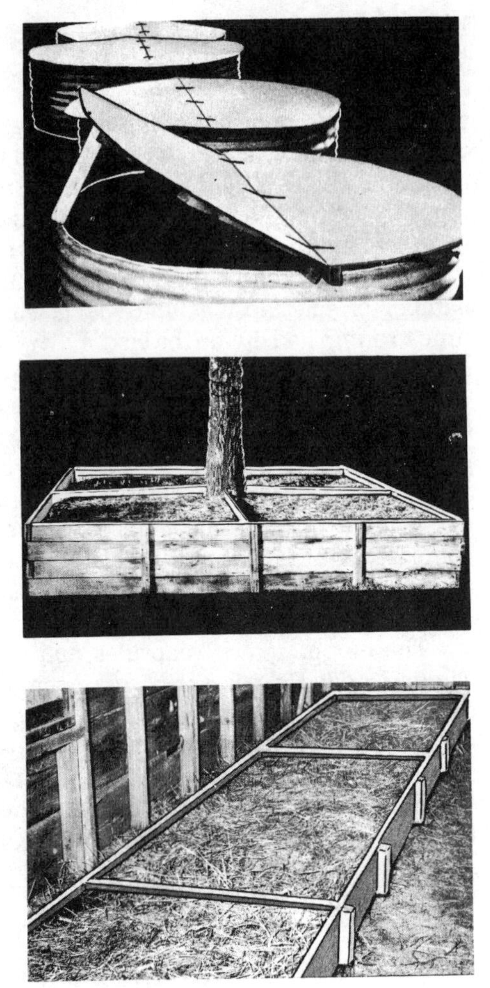

Building Indoor Pits and Bins

Courtesy Minerva E. Cutler, David's Folly Earthworm Farm

INDOOR earthworm pits may be built of sturdy, 2-inch rough lumber, as illustrated, to any convenient size . . . or they may be constructed of cement or cinder blocks, concrete, or any other materials that may be at hand. Wood pits are best built a few inches above the floor, both for convenience and for preservation of the lumber.

Indoor bins are used in most northern areas so that production can be maintained through the winter months. A barn or other building is often used. Heat has to be provided to keep the worms warm enough to continue production and to make sure they do not freeze during the cold weather. Gas wall heaters are sometimes used and should be vented to the outside. If they are vented inside the building, it could kill the worms.

BUILDING INDOOR PITS AND BINS

For a small operation, however, the grower may conduct his entire hatchery operation within the limits of a basement or other heated room. A good sized basement, fully utilized, can produce as many as a half million bait or breeder size worms per year or a greater number of pit-run.

Inside pits can double as garbage disposal units, a convenient way to dispose of kitchen wastes which are good food for the worms, supplemented with a little poultry mash or other ground grain additives. Garbage should be buried beneath the surface of the bedding, where the worms will soon dispose of it with no trace of odor.

Plans for constructing a unique indoor worm bed are available from H & K Research & Development Co., Inc., P.O. Box 44184, Omaha, NE 68144. It is ideal for those who want to raise worms indoors on a small scale.

Photo courtesy of H & K Research & Development Co., Inc. shows construction of a double bed which permits raising more worms in a limited space. It is neatly designed for those who want to raise worms indoors.

Another plan for indoor operation is illustrated by this photograph from Leach's Bait Farm in Kentucky. Construction is simple and inexpensive, using a wood framework covered with heavy plastic. 55 gallon oil drums were cut in half for the bins. A wood stove provides heat.

Inside bins, supported on legs, are shown in the photograph below from Albrecht Earthworm Hatchery. These bins will support 3,000 or more worms each. Note that the legs are set in cans, which contain a small amount of old oil, or kerosene, to keep ants out.

The procedure for indoor beds is identical with that for outdoor pits, so far as bedding, feeding and harvesting are concerned.

Wood bins should be treated with some preservative to lengthen the life of the wood. Creosote is not recommended because of its possible adverse effect on the worms, but a hot paraffin wax may be used, or melted pine tar pitch, either of which is harmless to the worms. Whatever the scope of your outside hatchery, it will pay you to maintain a few indoor winter bins.

Construction of Outdoor Pits

This photograph shows a typical above-ground pit, 48 inches wide by 72 feet long (inside measurements). It is built of cement or cinder blocks, two blocks high, all above ground; a suitable construction for a mild climate area. In cold areas, three-block-high construction is favored, with two blocks below the ground level for better cold weather protection.
John R. Pryor photos.

THERE are many practical methods and materials for building outdoor pits; concrete, cement or cinder blocks, hollow tile, brick or treated rough lumber. Discarded stock watering tanks, partially buried, makes good pits, as do oil barrels cut in half, or any other metal receptacle that is large enough for the purpose.

For convenience, rectangular pits should not be more than three feet wide, a practical width for working the pit from either side. The length can be suited to the space or materials available, from 6 feet to 100 or more. Long pits, however, are more convenient to work if divided by partitions into shorter segments, say 6 to 10 feet. You'll want to work your pits in rotation when you are ready to fill orders, and it is easier to work with units of medium size. A pit 3 x 20 feet will maintain from 50,000 to 60,000 mature worms, and with several turnovers per year should yield at least 200,000 worms annually.

Most breeders dig outdoor pits from 16 to 24 inches deep, for two or three blocks, if cement blocks are used, with one block above ground level to keep out surface water. The depth depends upon the climate. In mild climates a three-block pit is deep enough, and easier to work than a deeper pit.

Pits may also be built entirely above ground in mild climate areas; are a little more convenient to work, but lack the summer coolness of underground construction, and in the colder areas are harder to insulate against freezing in winter.

The same pit shown on the opposite page, filled with compost, ready for stocking. You will note in the opposite photo that a floor of boards, loosely spaced, has been laid over a shallow gravel fill, which provides better drainage. Boards prevent digging into the gravel when the pit is worked. For greater convenience, a long pit may be partitioned into a series of shorter beds.

Drainage should be provided by building in a screened opening in the floor of the pit or into the side walls a few inches above the floor, connected with field tile to carry off the excess water. A drain 3 or 4 inches above floor level is considered best. Some water in the bottom of the pit will do no harm; in fact some of the breeders maintain that worms do better and grow larger where there is water in the bottom of the pit.

A 3 to 5 inch layer of coarse sand and gravel in the bottom of the pit is helpful in taking care of excess water; or even better than gravel is a layer of crushed limestone, which helps to correct acidity in the pit. Over this layer of gravel or limestone lay a covering of boards, spaced a half inch or so apart. That's so you won't dig into the gravel when you work the pit, or when you empty it to renew the bedding.

Load the pit with about 12 inches of bedding and plant it with approximately 100 to 200 mature breeders per square foot of bed surface area, or 500 to 1,000 pit-run worms. When all worms have gone underground, wet it down and cover the entire surface with a layer of burlap or boards. That's all for the first three or four days. Then, after the worms are settled in their new environment, you may feed them. Remember to sprinkle the pits as often as necessary to keep the contents moist. DON'T FLOOD IT. More about compost and bedding in another chapter.

The initial filling of compost should be rather shallow, so you will have plenty of room to add new layers of compost from time to time to provide new food and bedding for the breeders.

This photograph shows a pit of plank construction, 6 x 60 feet, cross partitioned to make 15 beds 4 x 6 feet. Under favorable conditions it should produce ¾ of a million worms. It is partially recessed into the earth, but because of comparatively shallow construction will require heavy banking in cold weather. Planks will last longer if painted, or treated with a heavy coating of pine tar pitch. Photo courtesy of A. L. Race, a Missouri breeder.

BUILDING PITS IN NATURAL SHADE

IF YOU have a woods or grove available, where you can construct your pits in natural shade, you are in luck. It will eliminate the cost of building sheds; you won't have to burn lights over the pits; and in this close approach to a natural habitat the tendency to crawl or escape is reduced to a minimum.

Perry Gibson, who sends us these photographs of his South Carolina worm hatchery, tells us that he builds his pits of 4-inch cement blocks, three blocks deep, half under ground and half above . . . 32 inches wide inside, and as long as is convenient. He fills the pit with a 50-50 combination of manure and peat moss, plus five to ten percent of dried oak leaves, rotted until they have passed the heating stage. For a bed 50 ft. long he uses 20,000 mature breeders as foundation stock. He keeps

the bed just barely wet, and the material well loosened. In winter weather, however, he doesn't loosen the bedding very deeply, and he waters it lightly.

As supplementary feed he uses chicken mash, spread thinly so the bedding can be seen through it. When the feed is almost gone he loosens the bed, waters it and feeds again. A little brown sugar sprinkled over the bed now and then prevents the worms from crawling out.

Mr. Gibson places short boards across the pit at regular intervals, then longer boards lengthwise down the middle to hold them in place . . . a sort of roof framework. He then covers the entire pit with six foot lengths of extra heavy tar paper a few inches wider than the pit, weighted to keep it from blowing off and overlapped to keep out water. These covers are never removed except for feeding and watering, or for working the pits.

The rodent poison known as "d-CON" is put in any rat or mole holes discovered around the outside of the pits, and also in pans inside the pits. In using this or any other poison in or around the pits, be sure it does not get into the bedding. Anything that will kill the pests will also kill the worms, if it is put where they can eat it.

A GOOD TIP FOR BUILDING A LOW-COST CONCRETE PIT

Writes Perry J. Sherman, a cement finisher from Houston, Texas: "To build an inexpensive pit, first dig the pit and smooth the sides and bottom, then line the sides and bottom with galvanized metal lath, holding it in place by pushing galvanized nails through the mesh into the dirt walls. With a mixture of plain sand and cement, plaster on about ¾ or 1 inch of "mud" and you've "got it made". The worms can't burrow out, and it is cheaper even than wood, which is bound to decay." Thanks to you, Mr. Sherman. It sounds good, and you may have made a valuable contribution to the science of earthworm pit building.

PROTECTION FOR OUTDOOR PITS

IT IS IMPORTANT that outdoor pits be protected from the elements . . . from excessive heat and dashing rains . . . by some kind of roof, although it need not be elaborate or expensive. A rough pole shed with metal, board or slat roof will do the job as well as a more costly structure.

The illustration above (Pickwick Worm Gardens photo) shows such sheds, built of cedar poles, with open roof framing, which is covered with chicken wire. On top of the wire is a cover of feed sacks, sewed together and anchored to the wire. This allows rain water to drip through gently, which is the best kind of watering when Nature cooperates, and at the same time keeps the pits shaded and cool in summer. There is usually a cooling breeze under such sheds, even on the most sweltering days; good for the worms and good for the workers.

The beds beneath these sheds are also of wood, built of 2-inch planks, with board bottoms to keep out moles. The plank construction is naturally less durable than concrete, and some of the worms are lost through crawling, but the beds are so much cheaper to build that the grower can well afford to replace them when necessary. The worms seem to do better in the wood pits also, and in all probability most of those that crawl merely find their way to other beds.

Another way to cover pits, without building a roof over them, is to build frames over them for using hot bed sash . . . an excellent means of "forcing" earthworm activity and promoting a longer spring and fall breeding season. By thus taking advantage of the sun's heat, even in winter, greater production may be achieved. The same results may be gained, inexpensively, by making frames for the pits and covering them with the new heavier weight polyethylene now available. In the use of either glass or polyethylene, some provision should be made for shading the pits in extremely hot midsummer weather.

The all-steel shed framework shown above is rather expensive, but permanent. It is located in central Florida and houses an African Nightcrawler hatchery — Justison's Worm Farm. In Florida, where wood deteriorates rapidly, these beds are made of the more durable cypress.

For a small back-yard project, like the one pictured here, when good appearance as well as utility is a factor, a pole shed with metal roof serves the purpose nicely, but can be "dolled up" a bit with the addition of some inexpensive lath lattices and the planting of a few climbing vines.

Bedding for Earthworm Pits and Bins

THE standard formula for compost or bedding for earthworm pits used to be: "one-third manure, one-third topsoil and one-third peat moss." In the light of continued experiments and experience, however, earthworm breeders have found that topsoil in the bedding, although it does no harm, has little value for the worms and may as well be eliminated. Domesticated earthworms, usually the red worm, are not earth dwellers, but inhabit manure piles, compost heaps, or any accumulation of decayed and rotting animal or vegetable waste. These are their natural habitat, and the more closely we can approximate their natural environment in the preparation of pit materials, the better they will thrive.

The most nearly perfect BASIC compost, or bedding, is a 50-50 mixture of manure and peat moss . . . the manure to provide a supply of food, and the peat moss to keep the bedding loose and to hold the moisture the worms need. The manure may be cow, sheep, rabbit, horse or poultry manure in limited quantities, sufficiently aged to a point beyond the heating stage, but not old enough to be devoid of food value. Most barnyard manure will contain urine. This should be leached out by running water through the manure pile. It does not require a long washing but enough to produce a steady flow through the manure. You may put the manure in any container that will allow the water to drain off.

In mixing manure and peat moss be sure that the peat moss has been soaked for at least 24 hours (a shorter period if warm or hot water is used) so the fibers will be thoroughly impregnated with water. So treated, it will give off moisture slowly for days and keep the compost damp. The culture should then be mixed and sprinkled repeatedly until the material is thoroughly and evenly moistened all the way through, but not soggy wet. Be sure there is no "heat" in the mixture.

To this basic mixture can be added many other things upon which the worms feed . . . dried leaves and grass clippings, weed or flower stalks, waste materials from cider mills or from fish and poultry markets, breweries, groceries, canneries. Also kitchen and garden wastes; old bread, vegetable tops and parings, fruit skins, melon rinds, meat trimmings, beef suet, sour milk, bottle rinsings and dish water. Avoid excess quantities of salt or vinegar.

In addition to all of these waste materials (and many large breeders won't be bothered with them), the beds should be top-fed with finely ground grain feeds such as poultry mash, calf meal, pulverized corn, cotton seed meal or similar products. These are the additives that add size and girth, so important in growing good bait worms. Such feeds may be spread thinly on top of the bedding or buried in trenches through the middle or around the sides of the pits . . . never mixed into the bedding, where it may heat.

Any coarse materials used in the bedding, such as corn or weed stalks, hay, straw, paper, corrugated boxes or leaves, should be ground or chopped and composted before they go into the bedding. As your project grows, you may need tons of specially prepared compost on which you can draw for bed materials. It may be made in a pile or heap, much as garden compost is made, but may be used as soon as it has broken down and passed the heating stage. Earthworms can finish the job after it goes into the pits.

There are now on the market a number of concentrated "activator" products which contain a high concentration of soil bacteria, developed specifically for the rapid break-down of organic wastes. These activators, under such trade names as Activo, Actimus and others, are advertised in gardening publications, seed and nursery catalogs, or by garden supply and seed stores.

Through their use, sprinkled lightly over each thin layer of organic materials as they are added to the compost heap, it is possible to produce finished compost in as little as three to five weeks, or a little earlier for earthworm bedding use. The activators also attract earthworms, or work with earthworms added from your own breeding pits to further hasten the processing of your raw materials.

IF YOU CAN'T GET BARNYARD MANURE

If you find it impossible to get barnyard manure, or if you live in an area where the use of manure is prohibited, ground or hammermill paper also makes an excellent bedding material. This can be used instead of manure, mixed with peat moss. Corrugated paper boxes contain glue which the worms love, so it serves as both a bedding and food source.

Bagged manure, such as may be found in garden supply stores and feed stores, may be used. Be sure it has not been chemically treated. A stock yards product consisting of stock yard wastes that have been ground, dried and bagged . . . hides, hair, entrails, manure, hoofs and horns . . . is good bedding and food material if it is available. It is clean and practically odorless.

Earthworm Feeds and Feeding

TOO many thousands of words have been wasted, in this writer's humble opinion, on the subject of earthworm feeding . . . a confusion of learned but conflicting theories and conjectures, and much sheer nonsense, apparently designed to impress the breeder, but succeeding only in confusing him.

The feeding of earthworms is, in fact, utterly simple and uncomplicated. In their natural habitat they live happily and in the best of robust good health, in any old manure pile, compost heap, garbage dump, or any other collection of decaying vegetable or animal waste, and that's all it takes to keep them healthy and happy in an earthworm pit.

Give them plenty of manure . . . cow, sheep, rabbit, horse or poultry manure in limited quantities. That serves as both food and bedding. To it may be added a wide variety of other waste products, such as kitchen refuse, slaughtering wastes, or just about anything that has lived and died, if it is reduced to proportions that will permit the worms to consume it.

Manure used strictly for feeding should be fresh, while aged manure should be used in preparing bedding. The fresh manure is applied on top of the bed in a trench through the center of the bed, then watered and covered with a thin layer of bedding. If it gets too hot, the worms have room to back off.

Small growers can take the time to use these waste products; large growers are too busy to be bothered with them. They pour tons of feed into their worm pits; can't take the time to fuss with glass clippings, table scraps and coffee grounds.

Like the cattle grower or the poultry breeder, however, the worm grower wants his livestock to grow fast, and he wants it large and plump. To accomplish that he supplements the manure diet with ground grain feeds . . . poultry mash, calf meal, soy bean meal, fine ground corn meal, or any one of a dozen other additives, such as alfalfa meal, ground hay, paper pulp, cotton motes, or whatever is cheapest and most plentiful in the grower's immediate area.

Ground grain feeds, because of their tendency to create heat, should never be mixed directly into the bedding. They should be sprinkled in a thin layer over the top of the bedding.

Grain feeds should be ground to a very fine texture so it can be consumed easily by the worms. Any grain left in the bed will sour and create an acid bedding problem. This is also the reason why it is best to feed lightly every day or two, applying only enough feed to last between feedings. Since the feed must be kept moist for the worm to consume it, spray the bed with water after feeding.

In our experimental pits, when we dig them prior to feeding, we fork the bedding into a high ridge through the center, the length of the pit. We then feed cow manure in the trenches along the sides, wet down the feed, then cover it by spreading and leveling off the center ridge. We then water the entire pit through a protective covering of burlap bags.

Another method of controlled trench feeding is to build pit-long feeding baskets, using half inch mesh hardware cloth, 6 x 6 inches or thereabouts, and as long as needed. The feed for the entire pit is put into this basket trench and well watered. The worms will feed in and around the edges of it, but can back away from it when they have eaten their fill.

Air dried sewer sludge, easily obtained in most communities, is one of the best possible feeds for fattening worms. It should be allowed to "season", for several months if possible, before using it in the pits. It is then best used in trenches, or in lumps placed at intervals over the surface of the pit. Sludge is recommended for fattening pits primarily, but should be used sparingly, if at all, in breeding pits. It seems to slow up the breeding tempo, possibly because fat worms are sluggish.

It has been noted, also, that sludge fattening seems to be temporary. As soon as it is discontinued, the worms quickly revert to their former size. As a fattening agent for bait worms, however, it is ideal, and lasting enough to serve the angler's purpose.

Bedding should be changed every six months or so, as its food value becomes exhausted. That can be done, without much loss of worms, by feeding heavily to bring them to the top, then recovering most of them from the top two or three inches of the bedding. A more convenient way, if your project is expanding, is to divide the old pit in the middle. Fork half of the bedding, worms and all, into a new pit, spread out the old bedding in both the old and the new pits, then add fresh bedding to both of them to bring the compost up to the proper level.

FATTENING EARTHWORMS FOR BAIT

(Adapted, by permission, from the popular book, "Larger Red Worms", by George H. Holwager)

SEPARATE fattening pits are essential for the production of large bait worms. Breeder pits can be fed and watered heavily to produce size, but this will slow breeding. Evidently fat worms do not breed well. On the other hand, newly hatched worms do not seem to thrive in the soggy, heavily fed pits required for fattening; hence, separate pits for the two jobs.

Fattening pits should be deeper in the ground than breeder pits and

FATTENING PITS
Note depth in ground, shade trees and winter cover of boards on one, hay on another. Stream at back supplies water.

should be heavily shaded. They should also have free water space at the bottom, gained by taking off the drain at a height of about six inches from the bottom of the pit.

We filled our fattening pits with air dried sewer sludge and soaked it thoroughly. We added a thin layer of stable manure and then placed worms from the breeding pits, together with the compost and feed we dug up with them, around the sides of the pits. Two or three days later we piled fresh sewer sludge six inches deep down the center of the fattening pit, being careful not to cover the space around the edge where we had placed the young worms. The pit was then watered lightly with a spray. Two weeks later we began harvesting. When the worms began to thin out we repeated the process.

We do not know of any adequate substitute for sewer sludge for fattening pits. It puts on size as nothing else we have found, and is by far the cheapest feed available. Most city sewage plants will give it away for the hauling. It is practically odorless, and entirely free from any danger of infection.

Sludge should be allowed to air dry for a year before being used alone in the pits. Fresh sludge must be used with extreme care because it may contain enough sewer gas to kill worms if a pit is covered solid with it. But a small amount of fresh sludge, in a trench, will fatten worms faster than anything else we know.

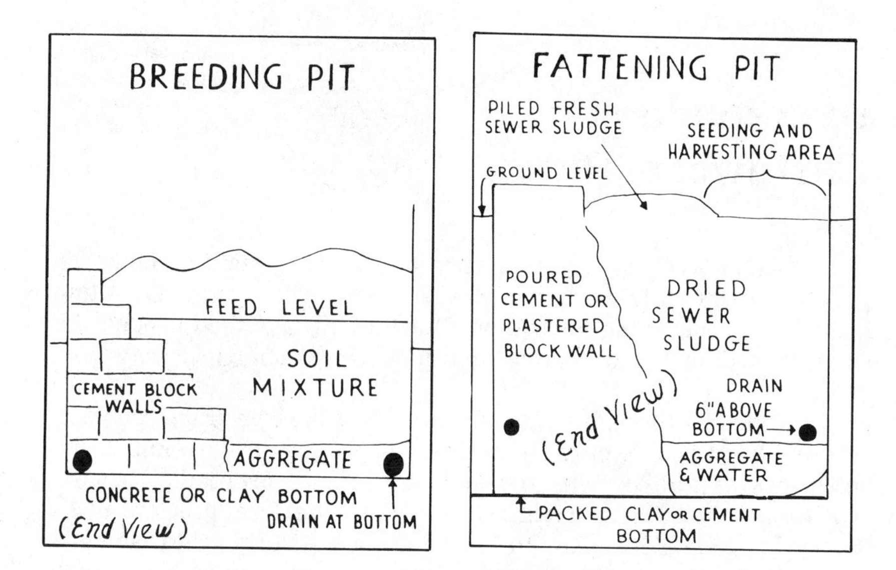

Holwager's DUAL-PIT method for breeding & fattening earthworms.

Over-acidity in Earthworm Pits

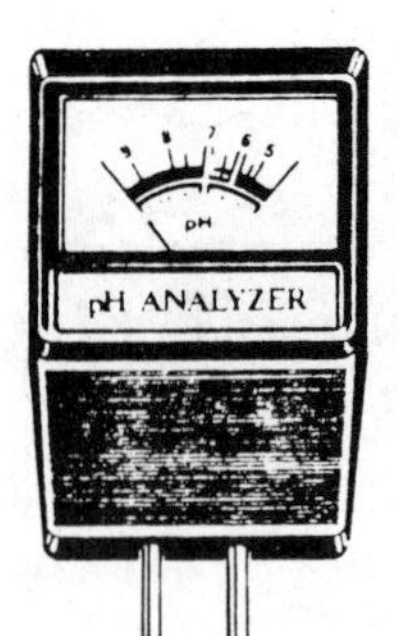

A pH meter. Probes are inserted into worm bedding to obtain the pH reading. Illustration courtesy of Environmental Concepts

EARTHWORMS require a certain amount of acidity, and in the natural state feed on highly acid materials, such as rotting manures and decaying vegetation. But they always feed around the edges of such materials, eating what they need, then back away from them.

Over-acidity throughout the pits, however, can be a grave problem, causing the worms to crawl in an effort to escape, or even killing the entire crop. Acidity is measured on the pH scale. A pH reading below 7 is on the acid side; above 7 it is on the alkaline side. Most growers agree that bedding should be close to 7 or neutral. A slightly acid reading is desirable. To test for the correct pH, litmus paper may be used and is generally obtainable from your druggist. Or you may want a more sophisticated pH meter which is available through a laboratory instrument supply company.

The best method for avoiding acid build-up is to avoid getting too much feed into the bedding. The ground grain feeds such as poultry mash will create some acidity and should be fed lightly. An acid bed with soured feed is an invitation to mites, often a real headache for the worm grower.

It is advisable to add powdered limestone (calcium carbonate) to the bedding on a regular schedule to counteract acidity. It may be sprinkled over the bedding and watered in, or it may be mixed with feed at the rate of one pound to each ten pounds of feed. The limestone seems to have no adverse effect on the worms, so it is not likely that too much will be used.

Where acidity has built up to the danger point before being detected, two immediate treatments of the pit are advised. First: give the bedding a heavy application of powdered calcium carbonate, spreading it evenly over the entire surface of the bed. Second: dig the bedding up into rough piles, allowing as much air as possible to get into the bedding. The pit should be kept moderately wet during this treatment. Give the calcium carbonate time to react. It is slow acting so it is possible to apply too much, causing the pH to swing to the alkaline side. This can create problems, too.

Protecting Earthworms from Their Natural Enemies

EARTHWORMS, like all wild creatures, have a host of natural enemies, such as centipedes, moles, ants, mites, rats and mice, gophers, toads, snakes and birds. Bred in captivity, they are easily protected from these enemies, but they must be protected, nevertheless.

Moles, for example, whose diet consists mainly of earthworms, can make serious inroads into your stock, if they can get into the pits before you know they're around. That's one of the principal reasons for building vermin-tight pits of concrete or of masonry blocks. If you have underground pits not so protected, keep a sharp lookout for moles and try to eradicate any that appear, by means of traps or poison.

Mice love to nest in the outdoor pits, burrowing down into the loose compost and dragging in soft materials for a cozy nest, where they winter in comfort, subsisting on the feeds intended for the worms. It is doubtful whether they actually destroy the worms; they are grain eating animals. But they are a nuisance. Most animals and birds can be kept out of the pits by screening them, but screens are troublesome and most growers won't be bothered with them.

Best cure for rats and mice is to keep a constant supply of the poison known as d-CON around the worm pits and nearby buildings. It is obtainable everywhere, and will rid the whole place of the rodents within two or three weeks.

One of the worst pests, so far as the earthworm breeder is concerned, is ants . . . not because they directly attack the worms (they don't), but because they rapidly rob the compost of sugars and fats, their favorite diet, which are also vital elements in the diet of the earthworms. Thus they can destroy or injure the earthworms, indirectly, by robbing them of important food elements in the bedding.

Ants may be killed or kept out of the beds by dusting the floor around the outside of the pits with pyrethrum, an insecticide containing roetenone. Be careful not to get it in the pits, because it will also kill the earthworms.

Mites of one kind or another are a constantly recurring annoyance to the earthworm grower. They are apt to be prevalent wherever there are ground grain feeds or rotting garbage in the pits. These, again, are harmless to the worms, but consume some of their feeds. Some, but not all of them, can be controlled by dusting the surface of the bedding lightly with sulphur or tobacco dust.

The red or "fishworm" mite is apt to become a real menace, and a difficult one to control. Most any household insect spray applied lightly to the surface of the bed can be effective. One of the more commonly used chemicals is Diazinon Insect Spray which is an Ortho product. It is used at a rate of two teaspoons per gallon of water which will treat several large worm beds. After application, do not water or feed the worms for several days.

Centipedes will destroy worms, but we know of no way of controlling them except to kill them as they are found in the pits. Fortunately they are not plentiful, as a rule, but you'll discover them from time to time as you work your pits.

One general bit of care will be useful in eliminating the CAUSE of many insect infestations. That is the liberal use of crushed limestone in the pits to keep the compost sweet. (Insect pests like it sour!) The use of limestone is covered in a separate chapter of this manual.

Growing Red Worms in Cold Areas

THERE is no area on the North American continent, from the Canal Zone to Alaska, too hot or too cold for the raising of red worms; but it is obvious that in northern areas, subject to sub-zero temperatures during the long winter months, some special rules for their protection must be observed.

Pits should be deeper than in the warmer areas, though two feet below the ground level should be sufficient if the pits are banked heavily on top and sides, and are not disturbed for the duration of the extreme cold weather.

Pits in the north are usually closed for the winter just before the first hard freeze, first having been fed heavily with enough organic matter to carry the worms through until spring. Such materials also create some heat. The worms are semi-dormant in winter and will not feed heavily.

The pits are left undisturbed until all danger of heavy freezes is past, and are then opened for the season's business, usually about the first or middle of April. The worms will be a little thin at this period, but healthy and lively, and with a week or two of heavy feeding will become plump and will resume their normal breeding activities.

Naturally the selling season is shorter in cold areas, unless provision is made for a few indoor pits, or heated inside winter storage, which most northern growers have. Some far northern growers depend upon heated buildings for their entire year 'round production, and as a result enjoy year 'round sales. Others keep only a limited supply indoors to meet the demand for winter bait.

When we asked a successful earthworm breeder of Toronto, Ontario, Canada, to give us an account of how she handled her earthworm pits in winter, she was good enough to comply with our request.

Her story follows, and we believe it will be of great help to all who raise earthworms in northern areas or plan to do so.

WINTER PROTECTION FOR EARTHWORMS IN COLD AREAS

THIS is the way we protect our worms, in outdoor pits, during our cold Canadian winters — a system that should work well in any part of Canada or in the northern United States, where sub-zero temperatures are common for long periods during the midwinter months.

"The pits or bins may be made of any one of the commonly used materials. We build our bins 2½ feet deep, with covers of wood, built in sections for ease in handling.

"Before freeze-up the bins have been filled level with compost and stocked with worms. Lengthwise through the center of this compost, to within 18 inches of each end of the bin, we dig a trench two feet deep by one foot wide. Using a mixture of chopped kitchen waste, fallen fruit and finely ground all-purpose laying mash, we spread a layer of this food up to 10 inches deep the full length of the trench, replace the compost and level off. The slow decaying of this mixture creates a gentle heat, and the worms will feed on it through the winter.

"Next comes the first winter blanket on the bins. For this, fresh (hot) horse manure is used. Mixed barnyard manure can also be used, but is not so good. It is most important that this first manure blanket be quite damp BEFORE it goes on the bins. Remove the wood covers from the bins, fork on the pre-dampened manure, spreading it out evenly all over the bin to a depth of two feet. Then using pre-dampened leaves, if available, bank them high and generously around all four walls of the pit to a height level with the top of the manure blanket. Replace the pit covers on top of the manure. It will seem high, but will settle.

"As freezing weather sets in, this over-all cover freezes hard, sealing the heat in and the cold out. It is important that this seal should not be broken; it could let in frost, and possibly damage the worms.

"The first week in February we spread a second blanket of fresh horse manure on top of the first one, approximately 18 inches deep. This time the manure is not pre-dampened. We replace the wood covers and let the new blanket settle.

"About mid-March, as the sun gets warmer each day, the dry second manure blanket begins to warm up, thawing the frozen one below. The leaf banking will thaw out a little more slowly.

"On warm, sunny days we remove the wood covers and let the sun shine directly on the manure, just an hour or so each day to speed up the warming of the bins.

"Soon the worms will have worked up near the top, and many can be sorted out from beneath the wood covers. We sort out the larger worms and put them into previously prepared fattening bins.

"We toss aside the leaf banking, when it is thawed out, together with the surplus manure. These, together with early spring weeds and lawn cuttings, are the basis for compost with which to set up our spring worm bins.

"The method outlined above for the winter protection of earthworms in outdoor bins is used with good results, in a locality where temperatures fall as low as 12-14 degrees below zero at times. In even colder areas the same method, with a greater depth of manure and leaves, should give ample protection from cold . . . and a good harvest of worms, come spring.

"It is not recommended that outdoor bins be opened before the spring thaw, but if a real need for worms arises, and the bin must be opened, the following method has proved least harmful to the worms in the bin:

"Choosing a mild, sunny mid-day, we remove the wood cover from the central part of one bin and cut around and through the frozen manure blanket. With a fork we remove and set aside the cut-out segment, taking care to break it as little as possible.

"In layers, lift out the compost until a concentration of worms is exposed. They will most likely be in the waste and feed mixture buried there in the late fall. Transfer worms and compost to a container and at once take them indoors or to a warm place. Now dump into this space any waste mixture ready at hand, replace the compost and level it off. Using care not to break it, put back the frozen section of "manure blanket" previously removed, fitting it in, as nearly as possible, just as it came out. Water around the edges of the break, so it will again freeze and seal the bin. Replace the wood covers, and snow, if any.

"To store worms for winter use we use metal laundry tubs, filled to within 3 inches of the top with a rich compost mixture. Each tub will store from 6,000 to 8,000 worms, all sizes. Over the top of the compost place a 2 inch layer of very damp garden soil, preferably clay type. Place the tubs where temperatures will remain at about 40 degrees. The worms will remain semi-dormant but will keep well for three or four months, needing no care. When needed, remove the soil topping, put the worms and compost into bins, feed and water. In a week the worms will be lively and normal.

"Several such tubs of worms were put up in the fall and what a gold mine they were around the month of February, when winter sales had lowered the supply of worms in basement pits."

Harvesting the Earthworm Crop

FOR maximum production, and minimum disturbance of the pits, you will find it advisable to work the pits in a regular rotation, giving them uniform rest periods during which the earthworms may feed, grow and breed without interference.

When pits are to be worked, they should be fed heavily on top the night before and, if possible, dug in early morning while they are feeding most actively. The feed brings them to the top of the compost, where you will find most of the worms in the top 4 to 6 inches so it won't be necessary to dig deep.

The worms should always be dug with a pronged fork, rather than with a shovel or spade, to avoid cutting or injuring them. Some diggers use garden spading forks, but the best tool, in our experience, is a long handled six or eight tined manure fork. The long handle saves the digger's back, the slender tines injure fewer worms, and the broad fork brings up a large mass of worms at each lift. Where a pit is "loaded", you may bring up as many as 1,500 mature worms in a single forkful.

Many breeders count the worms individually, or in lots of five or ten, in preparing them for the cartons. That sounds slow and laborious, but it isn't as slow or hard as it sounds. Pickers become surprisingly adept at this task, and many achieve a speed of 2,000 to 3,000 worms per hour, or even more, if a bed is heavily populated.

In one large bait growing hatchery, the pickers use carriers holding ten cans, much the same as the carriers used for berry boxes. They count 100 worms into each can (with a few extras), then return the carrier to the packing table for final packaging and boxing. One digger goes ahead of the pickers, turning up the worms for one or more pickers.

For one or two persons working alone, a good tool is a device like an ordinary sand screen, but with a solid rather than a screen bottom, set at a rather low pitch with the low end over the edge of the pit. This can be loaded from top to bottom with worm-filled compost, then the picker, starting at the bottom, fans out the compost and works it back into the pit, picking the desirable worms as he goes along.

For small lots, sheets of tin, or a platform of boards, may be laid across the pit and loaded with compost, preferably in sunlight, or with electric lights overhead. The light will drive the worms down, and the top dirt can be raked off into the pit.

After several such rakings, the worms will be in a solid mass at the bottom where they can be picked up by handfuls. With experience you will learn to pick up, say, 100 worms at each handful, which makes the handling fast. An extra small handful (worth less than the time saved) added to each thousand so counted will assure a fair and liberal count.

Another convenient method is to work on benches or tables of convenient height for standing or sitting, using large, flat, shallow trays for bringing worm-filled compost from the pit to the bench or table, as shown in the accompanying illustration. As each tray containing several hundred baits or breeder size worms is emptied, it is refilled or replaced by another tray.

Some growers, particularly in the West, sell worms by weight, at so much per pound, or by count established by weighing. To arrive at a reasonably accurate number of worms per pound they spot-check by counting an exact pound at regular intervals, or when changing from one pit to another, where the sizes of worms may vary. To compensate for inaccuracies in count, by this method, a few extra worms should be added; a bonus that is more than justified by the saving in time.

If the worms are being packaged for retail bait, count them directly into the containers, in 25's, 50's, etc.; or for bulk shipment, about 1,000 bait or breeder size worms to each gallon . . . pit-run, 1,500 to 2,000 per gallon, dependent upon size.

A recent development in the earthworm industry is the mechanical worm harvester. We have seen two styles, both incorporating screens through which bedding, capsules and small worms will pass but the larger worms will be retained and collected in harvesting trays or boxes.

One is a rotary type using a screen drum. The other is a flat oscillating screen. Both are motor driven. Bedding with heavy concentrations of worms is fed into the high end of the harvester and the screen motion and gravity move the material and worms along its length dropping the bedding, castings, and small worms through the screen into a container, and depositing larger worms at the end.

Various methods of "trapping" worms have been devised to avoid the necessity for digging them. Screen wire traps, or troughs, are buried in the bedding and fed heavily to attract the worms; then, after a suitable interval, the entire trap is lifted out with its heavy concentration of worms.

Morgan, in his book "Profitable Earthworm Farming", describes this method in detail and also includes instructions for grading the worms for size.

Shown here is a rotary type harvester in operation.
Photo courtesy of Cyclone Worm Harvesters.

There has been some experimentation with the harvesting of capsules. Storing and shipping them, of course, requires some know-how. According to David M. Davia in California, he has perfected a capsule harvester and has successfully shipped capsules to Europe. In connection with his capsule harvester, he has developed some worm raising techniques which he says produce large worms in a short period of time. This is apparently due to frequent removal of capsules and castings, thus preventing overcrowding by new worms.

Packing and Shipping Earthworms

BREEDERS employ various types of containers for packaging earthworms for shipment. Waxed ice cream cups and cartons with tight fitting lids, punched to admit air, are most commonly used. They are available in half pint, pint, quart, half gallon and one gallon sizes, and are the least expensive if used plain, unprinted and unlabeled.

More and more growers, however, are coming to the use of the cartons specially designed for earthworms, attractively printed and with covers perforated for air. The alternative, of course, is to have an individual label designed and printed for use on plain cartons. Under either of these plans the cost is a little more, but the advertising and sales value of such attractive cartons makes the extra cost worth while.

"Breather" type paper bags, specially designed for worm packaging and shipment, have become popular. They are light, efficient and relatively inexpensive. They are made in a wide range of sizes, half-pint to one gallon capacities. There are several suppliers of these bags. Refer to the latest edition of the Earthworm Buyer's Guide.

The cylindrical printed cartons and cups illustrated are manufactured by Sealright Co., Inc., 2925 Fairfax Road, Kansas City, Kansas 66115, who will be glad to supply price information and address of the distributor nearest you.

A half-pint carton will carry 50 bait size worms; a pint, 100 or more; a quart, approximately 200. For bulk shipping, gallon cartons are usually used, with 1,000 mature or bait size worms to the gallon . . . 1,500 to 2,000 pit-run worms. They will arrive in better condition if they are not crowded.

The standard packing material is peat moss, pre-soaked for at least 24 hours and squeezed fairly dry . . . dry enough so that a handful of it, if squeezed hard, will yield no more than a few drops of water. This is particularly important in hot weather. If too much water is left in the peat moss the worms will scald and die, and will arrive a stinking mess! Fairly dry peat moss will carry the worms through in good shape, even over a long haul. The peat will hold sufficient moisture in its fibers, and the worms themselves will provide some moisture. Peat moss has the further advantage of being light in weight, which cuts down postage bills or express charges.

Most growers use either parcel post or United Parcel Service for fast, economical transportation of worms. All earthworm shipments should be marked, by means of a label or stamp: "LIVE EARTHWORMS . . . DO NOT EXPOSE TO EXTREME HEAT OR COLD . . . OUTSIDE MAIL." Mail clerks are accustomed to earthworm shipments and are usually conscientious about protecting them.

It is not necessary, as a rule, to add worm food to the cartons of worms packed for shipment. For long distance shipment you may, if you choose, put a small amount of fine corn meal or chicken mash in the top of each package. The worms, however, will travel long distances without food of any kind. They may get a bit thin in shipping, but will pick up again rapidly when they are bedded down and fed.

For outside shipping cartons, corrugated boxes in various sizes may be purchased from any box manufacturer or jobber in large quantities, or if your needs are more limited we suggest that you gather up the cartons thrown out by grocers, hardware dealers, druggists and other merchants. They will cost you nothing, and will save the merchants the cost of having them burned or hauled away.

Boxes should be securely taped for shipping. You may use an extra strong filament tape obtainable at almost any stationery store. It is rather expensive, but it makes a neat, light, strong and safe package.

LOW-COST WORM SHIPPER

THIS shipping container was brought to our attention by a southwestern grower who developed it for his own use. It has a number of interesting features. It eliminates the need for outside shipping cartons, which means a savings in money, time and labor.

PACKING AND SHIPPING EARTHWORMS

The container is constructed of corrugated board, waxed heavily on the inside. There are die cut air holes, and it has the words LIVE EARTHWORMS printed on two sides. It is a sturdy light-weight (5 oz.) mailable container that is easy to handle.

It is designed to accommodate 1,000 bait or breeder worms, or approximately 2,000 pit-run worms. For larger shipments, you can tape or tie several containers together or place them in a larger outside shipping carton.

The containers are shipped flat and require little storage space. To assemble, you need only regular packing tape.

PUNCHING DEVICE FOR PERFORATING CARTON COVERS

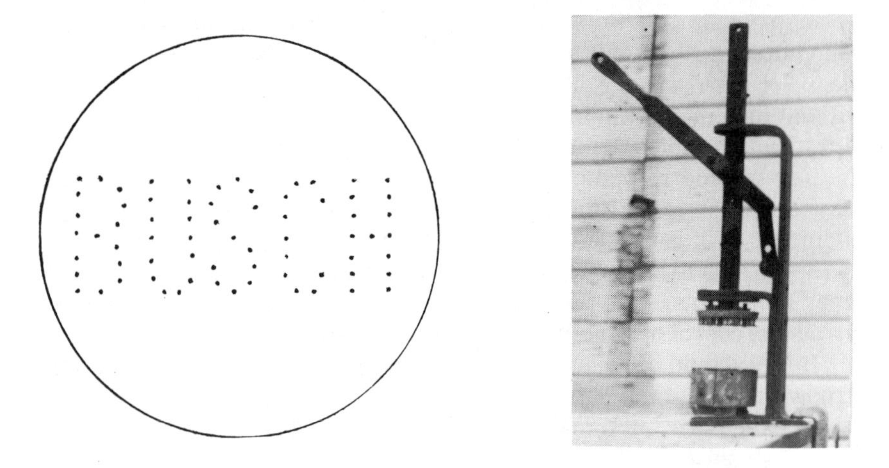

FREQUENTLY we receive inquiries from worm growers as to whether there is a faster or more convenient way of perforating carton covers than by just stabbing them with an ice pick. We don't blame them; it's a tedious job.

J. H. Busch of Milton-Freewater, Oregon, told us of his one-movement punch press pictured here. He made it of strap iron. The machine, which may be clamped to a table, will perforate covers of any size, and also flat discs, in a single motion. The spikes in the punching disc may be arranged to form a specified pattern or name, if desired. It is easy to operate, and because of its solid metal construction, there is little that can go wrong with it. It is an idea you may want to adopt.

Postal Regulations on Earthworms

EARTHWORMS may be mailed at regular parcel post rates, and in most instances special services, such as insurance, special handling, COD and air parcel post are available. They may be fully insured if that is desirable, and may be sent "Special Handling" to speed delivery. Insurance is costly, however, and most of the growers prefer to assume the risk of loss (which is infrequent) rather than pay the insurance fees.

Packages should be securely taped with filament tape, so there is no possible chance of breakage, and should be plainly labeled as to the contents . . . also marked OUTSIDE MAIL, so packages will not be put into regular mail sacks. The sacks are pretty nearly air tight and bad for the worms. Also, if a worm package should break, it could damage other mail.

As a rule, outer shipping cartons are not tight enough to require the punching of holes for air, but it is a wise precaution to punch a few holes anyway. Corrugated cartons let some air through, however, and worms require very little of it.

If there is any question as to rules and regulations governing the shipping of worms from your post office, consult your postmaster. If he isn't sure, he can get a regional office ruling. Packages may weigh up to 70 pounds which is more than enough leeway. Too much weight will damage the worms. One way to help prevent damage of this sort is to package the worms in several bait bags which are then put into a larger shipping carton. It helps prevent smothering. This could be compared to an egg carton with partitions. Consult your postmaster for package size limits.

A new kind of farming

There's money in angleworms

By BILL AKERS

There are as many kinds of

had its real start back in the 1950s and has been growing since. It is

you will have 1¼ million worms and egg capsules.

business they both would like to sell the mobile home park and

The Warrens do their harvesting on a table top

able to sell $1,400 worth.

Warren and his wife are still in the mobile home court business — in Soquel now — but they are also in the worm business in a big way.

On his land across from the court is his worm operation — a very neat, clean and odorless setup consisting of worm beds, a couple of small buildings in which his tools, equipment and supplies are housed, and tables on which the worms are

down. This process is repeated until all of the compost is separated and all that remains are the worms.

Worms-an easy crop

By KARL SIMON
Copley News Service

LOS ANGELES — They wiggle and wriggle. They squirm and they squiggle.

To young boys they're a delight. To little girls they're a fright.

But to Ivan Rayworth, the lowly earthworm is a crawling

gives one the impression Rayworth is a factory catering to the extra tall.

developed an interest in worms
them in his backyard.

regret is that he didn't get in

I would be supreme in the busin

are the mainstay of his b

, Rayworth says, bu
lifetime supply.

The Ruby Worm Farm ships worms postpaid for $10.50 per thousand.

Oak, elm and maple leaves, mixed with peat moss and sawdust grow salable earthworms for this Missouri into a fine complete organic compost.

The worm queen of Santa Ynez

By Irving Townsend

SANTA YNEZ — Nearly two years ago Alicia Landon switched from ponies to worms because she had heard that worms were profitable and she knew that ponies were not. Since then she and her next-door neighbor and partner, Charles Silva, have had to overcome bad advice, live through 18 months of trial and error, including dietary experiments, the creation of proper living quarters and escapes, devote long hours of hard work, face

The Early Bird Gets You-Know-What

By ORVILLE

exchange and furnishes sundries.

When relaxing, Hughes either listens to jazz or thumps out a ragtime tune on his piano or recalls earlier days in burlesque.

He started raising worms for his own use, because he liked to fish in the Tennessee River, on whose banks Savannah stands.

Fishermen started asking him where he got the fine worms he was using. He told them he raised them. "They started asking me if they could buy some, and as the requests increased, I increased the number of worm beds," Hughes said.

At first, he fed his worms grease and coffee grounds but found that drew insects. "Besides, there were

THE WALL STREET JOURNAL.

Did You Ever Wonder Where All the Worms For Bait Come From?

Here's Answer: Worm Farms; Red Golds & Red Wrigglers Make Rabbits Seem Sluggish

By STEPHEN J. AINSWORTH
Staff Reporter of THE WALL STREET JOURNAL

DECATUR, Ill.—Charlie Marsh, a construction worker turned entrepreneur, shows a visitor around his farm. It covers a quarter-acre, and the structure is a weathered shed with

SAN JOSE, CALIF., MONDAY MORNING, APRIL

Lil' Red Wrigglers Be
Than Lure Of Televi

By JIM BROADY
Mercury Staff Writer

What is long and skinny,

cial effects on inc plants.

And since the C

BUSINESS BURGEONS —Dave Shaver makes an was bitten by the worm bug.

Under ideal conditions, the worms can double their number in 90 days, but cold

Worms: How to get rid of your garbage and create a business

FLOWERFIELD — First there was Black Power. Then came Chicano Power, Indian Power, Women Power, and with the energy crisis, the possibility of no power.

Mary Appelhof, a resident of the unincorporated village of Flowerfield, Rt. 1, Three Rivers, has a new one to add to the list—worm power.

Worms, she claims, not only make good fish bait, they have "great capabilities for soil improvement," help speed up the decomposing process in compost piles, and also make a good garbage disposal.

Miss Appelhof and her two partners, Ilda Wissman and Bernie Zivanovic, are now in the worm growing business.

Starting with 2,000 worms two years ago, Flowerfield Enterprises, as their business is called, sold 500,000 worms (give or take a few) this past season, and this year is aiming for sales of 10 times that.

The worms go directly to fisherman, to bait shops, gardeners, and as "breeding stock" to other persons going into the worm business.

One entrepreneur they got

WORM POWER — In the left photo, Mary Appelhof finds some that she pulled out of the large bin at right which serves

of humus which makes good planting soil. Miss Appelhof estimates the 5-10,000 worms in the bin can take care of about pounds of garbage per week. Surprisingly, the bin emits any odor.

A by-product of the worms their way through the

THE CHRISTIAN SCIENCE MONITOR

An International Daily Newspaper

Focus *on worm power*

By Peter Tonge

Boston

He Makes $600,000 a Year Selling Worms

By CHARLES GOLDEN

His friends call him "Wormy." But Fred Rhyme can afford to laugh because he sells 250 million worms a year and makes $600,000 doing it.

"When people needle me about living with worms," he said, "I tell them, 'They may be worms to you, but they're bread and butter to me.'"

Rhyme began what he describes as

WIGGLY WEALTH: Fred Rhyn... checks some of the 250 milli... worms he'll sell this year.

was Rainbow Mealworms born.

Rhyme sells some of his worms ... holds others back to go through ... 3-month cycle to replenish his su... At present, he has between 100 an... million worms in the hatching ... and about 20 million ready for

Worm Farms Big Business For Two Cedar Hill Firms

Robert Williams, the undisputed worm king of the southwest, has taken a project that few people understand and made it into a worthwhile business.

In January of 1955 Williams started raising rabbits and put worms under the hutches to help keep it more sanitary. It wasn't too long before he realized that the worms had a better future than the rabbit business.

In 1957 the business had grown to such an extent that Williams had to retire from his position as printer at the Dallas Times Herald so that he could devote more time to his worm business.

Mrs. Jolinda Williams, who has been a part of the business from the beginning said that the Oakhaven Farms probably has the largest mail order worm business in the United States and that they have mailed worms to ... as Canada ... foundland.

The wor... haven ar... Hybrid, a... only by ... easily rai... for.

The wo... ... and are w... ...ared for

Mrs. ... worms ar... will multi... times in o...

The Wil... 1/4 acre o...

Miles I. ... Dorothy w... haven in ... ing the w... them for ...

At the ... year, fro... Walter P... to help.

Mrs. Z... Mrs. Will... in the ... volume of ... Oakhaven

Custom... breeding ... may tak... special c...

writing for information or help with any problem encountered in their new business.

Williams has written a book called "How to Sell Fishworms By Mail" and thousands of copies of the book have been sold.

The worms are sold by the thousands and the smallest amount mailed is 500.

Mrs. Williams said that during the first three months of the year sometimes Oak... have n... will mail millions of worms.

The c... from ad... 100 ma...

Harol... (C...

WORMS CATCH CASH

Take advantage of a business which is profitable anywhere in the United States and which requires little money to start

PRICELESS EARTHWORM PUBLICITY

Newspaper and magazine stories like these are worth their weight in diamonds! They have catapulted many a grower into the "Big-Time". For more data on getting such a story published on your own hatchery, see the chapter on Free Publicity, elsewhere in this book.

WORM FARMERS—Bill Holbert, foreground, and Ron Gaddie claim there's money to be made in worm... Times photo by Michael Ma...

SUCCESS IN THE SOIL

Tale of Two Worm Farmers

BY MICHAEL SEILER
Times Staff Writer

This is a sto...

How a BIG Worm Grower Operates

TED GALLAHER'S Pickwick Worm Gardens may not have been the world's largest earthworm hatchery, but it ranks high, not only in size but for steady, uninterrupted growth and for sound business management. Formerly the owner of an impressive chain of successful dry goods stores in southern Tennessee, Mr. Gallaher grew earthworms because he liked the business and the healthful outdoor life that it affords.

He modestly admitted an annual sales volume of forty to fifty thousand dollars, a payroll of twenty thousand dollars or so, twenty or twenty-five regular employees through the eight months' "busy season", with a skeleton force through the winter, and everything else in proportion.

Pickwick actually consisted of two large hatcheries, either of which would seem to entail a full-time management job. The customers were large retail bait dealers from coast to coast, most of whom ordered in large quantities, from 10,000 to 50,000 at a time . . . sometimes by telephone at 4 o'clock in the morning! Ted said it's a "dog's life", but there was a twinkle in his eye when he said it!

For the statistical department: 750,000 worms per week in peak weeks; 3,000,000 in one peak month; 700,000 in one outstanding single day. Two trucks operated between the hatcheries and the post office, one large and one small. One or the other made the trip daily. The accompanying photograph shows the small truck, loaded with more than 200,000 worms, 5,000 to each carton. The big truck handled 300,000 or more.

Pickwick worm pits, banked and fed for winter.

The pits under this shed, each 70 feet long, have been given a heavy feeding of crushed corn along the side walls, and are banked high through the middle with cotton and new compost, which will keep the worms warm in case of a freeze. The dark areas near the outer walls indicate where the worms are eating from the outside toward the middle. You will note, too, that plenty of space has been left in the pits for adding new compost.

The lights that you see burning over the pits are considered a "must" to prevent crawling, particularly on rainy nights or on damp, dark days.

The crushed corn which comprises winter feed for the worms is ground or crushed right on the premises . . . corn, cob, husks and all. (See illustration.) This is the waste from a corn shelling operation; a cheap feed that also serves to keep the beds warm. It is placed on the beds pretty thick and left all winter, renewed as often as the worms consume it. The pits in this particular shed require about three tons per month. It is kept moist, and by spring is sufficiently decayed so it can be turned into the beds.

In the event of a long, hard freeze the cotton and compost piled high in the middle of the pit will be spread out over the entire surface, taking care, however, that it does not get too hot.

GRINDING WORM FEED AND BEDDING ON A LARGE SCALE

ONLY in a very large hatchery where many tons of feed and bedding are required each month would a grinder of this capacity be required. It is a Letz Silage Mill for grinding corn, fodder, hay, paper and all kinds of roughage, mostly for bedding material. Different size screens make it possible to crush or grind to any degree of fineness required.

Pickwick Worm Gardens were the first (so far as they know) to make use of ground paper for worm feed and bedding on a large scale . . . corrugated board packing boxes plus all other kinds of waste paper . . . all wood fiber products that make excellent earthworm fodder. In the beginning, when Gallaher also operated a chain of dry goods stores, those stores provided tons of such paper, sufficient for the entire earthworm operation.

The ground paper was blown over into a tall roofless building adjoining the grinder room, open to rain and weather, where it could rot down naturally with no special treatment of any kind. The compost so formed was then fed from the bottom of the pile.

In later years, having sold the dry goods stores, and with the earthworm operation having far outgrown any possible local paper supply, Pickwick turned to other materials to obtain the necessary volume. Paper was still used when available, but the process of breaking up the boxes and putting them through the mill was slow and time consuming, an important factor in a large operation.

For smaller hatcheries, efficient smaller grinders are available at reasonable cost, designed primarily for compost grinding and shredding. The advertising of several manufacturers of small grinders appears regularly in ***Organic Gardening,*** Emmaus, Pennsylvania. You may easily obtain a copy, if you are not a subscriber, or you may write to the publishers, Rodale Publishing Company, Emmaus, PA.

SOME OTHER TYPICAL LARGE WORM FARM OPERATIONS

Well shaded concrete pits at a large Tennessee farm. The packing and shipping department handles the large volume of orders. Courtesy Hughes Worm Farm.

Used compost and worm casting are piled high at the end of a conveyor for easy removal from the building. Courtesy Ruby Worm Farm.

Earthworm Advertising and Selling

NO MATTER how fine the earthworms you raise; no matter how excellent your service; the world will never beat a path to your door unless you let the world know what you have to offer it. And your hatchery will never grow BIG unless you step out beyond the limits of your local community.

The U.S.A. is your market for earthworms. Possibly 80 to 90 percent of all earthworm sales are made by mail. The remaining 10 to 20 percent is sold directly from the pits to local customers in the area where they are grown. The growers who advertise, who sell from coast to coast, are the ones who grow big. Most of their customers they never meet personally, but through good worms and good service they have cemented the friendship of those customers and have gained their complete confidence.

You don't have to be a large grower to take advantage of the big national market. You can tailor your advertising to the volume of worms you have to sell, possibly only one or two publications in the beginning, and increase it as your earthworm volume increases. Over the long pull you will find that national advertising costs you less per unit of sale than does local advertising and the leg-work that goes with it. Postage stamps are much cheaper than gasoline, the wear and tear on your car or truck, and the many hours of your own time involved in local selling.

Please don't misunderstand these observations. There's nothing wrong with local selling. It is desirable business, and you are right in obtaining all of it you can secure. But if you have the urge to grow big you will want to supplement it with mail orders sales, avoiding the limitations and sometimes the competition of local business.

To be sure, there is also plenty of competition in mail order selling. You have only to scan the classified columns of the outdoor sports magazines to observe it. But the market is so huge as to seem almost inexhaustible. Millions of readers for every ad that appears; millions of fishermen; thousands of bait dealers; thousands of readers who are potential earthworm raisers, and therefore potential breeding stock buyers . . . there is no end to the demand that exists, or can be created.

Naturally, advertising and sales promotion methods demand an investment . . . an investment, however, which will return to you with interest if it is made wisely and cautiously, and which with repeated turn-overs will begin to "snowball" — to increase in efficiency and in cash returns.

FREE PUBLICITY FOR YOUR HATCHERY

ONE of the sweetest forms of "advertising", and one that will give your business a quicker and more substantial boost than even your best paid advertising, is one that has no "price tag" on it . . . the free publicity that worm growers frequently receive through the newspaper sports columns, local publication news stories, and occasionally feature stories in outdoor sports publications or national magazines. (See reproductions of some of these stories in the center spread pages of this book.)

Worm raising as a business is still novel enough to make it newsworthy for many types of publications, and editors or free lance writers love to uncover really good story material of this kind . . . particularly "success" stories about persons or families who, from small starts, have built profitable businesses.

If you are sufficiently gifted in the art of writing to do a good human-interest story about your own earthworm project, by all means do so, and send the story to any sports editor or magazine editor to whom you think it might appeal. If it doesn't click, don't be discouraged. Send it out repeatedly (but to only one publisher at a time) until it is used. Don't worry about getting paid for it; the favorable publicity is reward enough.

A better method, and one more acceptable to publishers, is to first send along a brief outline of the story you have in mind, so the editors may advise you in advance whether or not such a story would be welcomed.

If you don't feel that you can do the finished story, send along the outline of it, or the essential facts, with the suggestion that "there might be a good story here" for one of the publication's staff of feature writers.

Whether or not you ever hit the "big time" in a press release or human-interest story, you can do yourself a world of good, businesswise, by talking about your project to local publishers, tourist court operators, fishing camp owners, sporting goods dealers, and others in position to send customers to you.

LOCAL ADVERTISING AND SELLING

IF YOU are located in a good fishing or resort area, where you feel that the local fishermen and bait dealers will use all of the earthworms you can raise, then you can concentrate on strictly local advertising, using the local newspapers regularly, plus plenty of good signs pointing the way to your place of business. You can't buy such signs ready-made. The best bet is to have signs made by a local sign painter.

All outdoor signs should be water-proof, of course, and done in color to attract attention. Lettering in Scotchlite tape, to reflect in the illumination of car lights, is an advantage, providing both day and night advertising.

If you distribute through dealers . . . tackle shops, filling stations, fishing camps, boat docks and similar outlets . . . it is advisable to have some good display cards made for counter and window use. Your local sign man can make them at modest cost. Put these up yourself to make sure they are used.

You will, of course, want to call personally and frequently on all of your dealers. You may "run a route", making your own deliveries to these outlets at regular intervals, keeping them well supplied, either on a cash sale basis or on consignment, collecting at each trip for the worms that have been sold since your last call. Consignment is not recommended if cash sales are possible. However, you may encourage outright cash orders by giving the buyer a little better discount for cash than for consigned merchandise. In any case, your arrangement should provide for taking back any bait that may have gone bad, and replacing it with fresh stock. The usual discount for outright purchase is 40% off the retail price, 30% on consigned stock in which the retailer has no investment.

In addition to your wholesale outlets, you will probably want to maintain a retail stand at your place of business. This is profitable but it may mean long hours, including Sundays and holidays. To avoid some of the inconvenience of filling orders at all hours of the day and night you might consider a self-service arrangement or, if the volume justifies it, the purchase of an earthworm vending machine. (See chapter on that subject.)

Compile a mailing list of all the wholesale outlets in your area and use it at key intervals . . . prior to the beginning of the season, or during it; a "thank you" card at the close of the season; a Christmas card at holiday time; anything, at any time, to keep in touch and cultivate good will. It helps!

NATIONAL MAIL ORDER SELLING

IN REACHING out for the national mail-order market . . . and it IS the big volume market . . . you will need an assortment of supplies not usually required in local selling. In mail order selling, since there can be no personal contact, your literature has to do your selling job for you, and it must be GOOD. More on that a little later.

In addition to sales literature you will need some office supplies. . . letterheads, envelopes, price lists and order blanks, instruction sheets for beginning breeders who may order foundation stock from you, and some kind of printed or multigraphed folder for use as an enclosure in answering inquiries or in mailings to general prospect lists.

We believe it will pay you, also, though it is not absolutely essential, to have a neatly printed label for your shipping cartons, and also wrap-around printed labels for your individual worm cartons. This last applies to either national or local selling. Such a label, featuring your own hatchery or brand name, your address, and possibly an interesting illustration, is not only good advertising but it lends an atmosphere of importance and business stability.

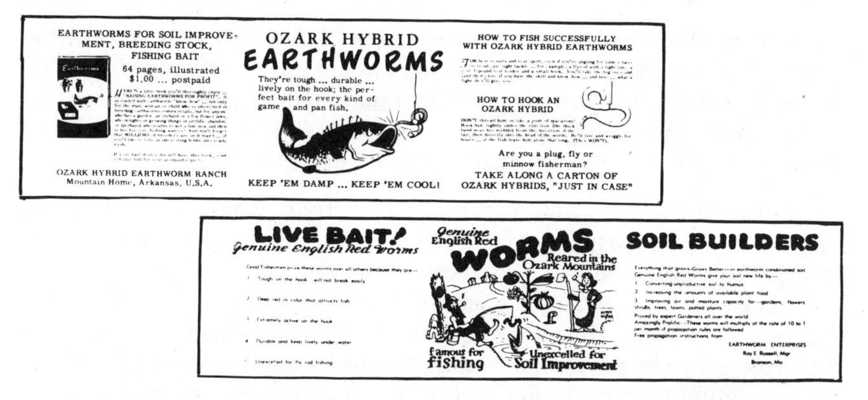

On such labels, and on your business stationery and office supplies, your local printer should be willing and able to give you a good deal of help, not only in the printing but also in layout and design, and the ordering of art work, cuts, etc. There are no stock materials available for these items, so far as we know. Individual requirements vary too widely to make standardized designs practical.

The shipping labels reproduced, suggest the possibilities of unusual label design. They can not, of course, be copied or reproduced, because the designs belong to the growers who use them, but they may help you in "dreaming up" an original design of your own.

Coming back to the subject of your key piece of advertising and

sales literature which every grower should have, you may for a while "get by" with a simple instruction folio made up of a few multigraphed or mimeographed pages, stapled together, which will answer in advance most of the questions from prospective buyers that you couldn't possibly find time to handle by personal correspondence.

More attractive, and far more effective, would be a fully illustrated folder, featuring views of your own hatchery if possible, and telling the whole story of earthworm care and culture. The big obstacle, of course, is the fact that such a folder is expensive, and in the beginning, at least, may seem out of reach. It is often true, also, that the grower has no particular skill or talent for writing such a folder, or for arranging it attractively, and professional talent for such work is highly paid.

There is an answer to that problem, however, and one that may serve you well. . . an entire series of advertising and sales promotion materials, skillfully written and attractively illustrated, printed in large quantities for economy and available in small lots to individual growers at modest cost. These materials consist of folders, with space for your imprint, low enough in cost so you can afford to give them away. . . and booklets, also with imprint space, which you can offer for sale at a very liberal profit to yourself. You'll find these materials listed in another chapter.

Outdoor Life

REG. U.S. PAT. OFF.

50 LIVE BAIT

"RAISING ... worms ... rofit", exciting, fu... 8-page manual. Redwo... can Nightcrawlers, everything for successful growing, advertising, marketing — postpaid. Shields Publications.

GOLDEN Mealworms (Extra Large). 500 - $2.00; 1,000 - $3.25; 3,000 - $7.00; 5,000 - $10.00; 10,000 - $19.00. Guaranteed Live Delivery, Postpaid. World's Largest Growers. Rainbow Bait, P. O. Box 4907, Compton, CA 90220.

SUPER Large Redworms, 100 - $2.00; 500 - $4.50; 1,000 - $7.00; 2,000 - $13.00. Postpaid with Instructions. G&R Farm, 1205-C Melroseway, Vista, CA 92083.

OUTSTANDING Foundation Stock (bait - soil - breeding) Booklet 50c. Braby's Earthworm Farm, 23190 Lyon, San Jacinto, CA 92383.

REDWORMS, Hybrid Select, Hand Picked, Generous overcount, live delivery guaranteed. Unlimited supply, shipped postpaid. Dealers, jobbers invited. $5.75 - 1,000. Write now for quantity discounts. Interested In Growing? Write now for complete raising, harvesting instructions to: Rocky's Live Bait Farm, 21126-A Bryant, Canoga Park, CA 91304.

REDWORMS, $5.00 - 1,000; $21.00 - 5,000. Robbies Bait, 11972 Lakeside, Lakeside, CA 92040.

LARGE Red Wigglers. Fishing, Gardening, Breeding, 1,000 - $6.00; 5,000 - $25.00; 10,000 - $45.00. Bedrun: 1,000 - $4.25; 5,000 - $15.00; 10,000 - $30.00. Ko-Part Worm Farm, Rte. 2, Box 275-A, Americus, GA 31709.

NEW Book - How To Grow Northern Nightcrawlers indoors or outdoors. Every question and answer. 25 years experience. $5.00. Satisfaction guaranteed. Ellis Lake, Alexandria Bay, NY 13607.

WORMS Redgold Hybrid. 1,000 - $8.00. Satisfaction. G&H Ranch, Box 103, Reese, MI 48757.

GRAY Crickets: breeders our specialty. Scientifically raised, vitamin fed. Heavy producers. Hard to get items now available. Write for supply catalog: Selph's Cricket Ranch "Midsouth's First—World's Largest." Box 2123. Memphis, TN 38101.

BAIT Dealers, Jobbers wanted - Nightcrawlers, Redworms, Supplies. Immediate shipment, Air, Bus, UPS. Wholesale Bait Company, Hamilton, OH 45015. (513) 863-2380.

CORN GRUBS-LARGE WHITE LARVA 200 - $1.50, 500 - $3.00; 1,000 - $5.00 Postpaid. Dealers prices sent on request. Dobosco, Box 2001, Hamilton, OH 45015.

REDGOLD Hybrid Redworms. 5,000 - $15.00. 10,000 - $27.00. Postpaid. Generous Overcount. Hillard Robbins, Quebeck, TN 38579.

RED Wigglers: Bait or Breeders. 1,000 - $5.00; 2,000 - $9.50 Postpaid. Bert's Bait Farm, Irvine, KY 40336.

LARGE Red Worms. Details, Box 998, Route 12, Greensboro, NC 27406.

HIGHEST Quality, Handpicked, Hybrid Fishworms. Millions. Bulk: 1,000 - $8.00; 2,000, 3,000, 4,000 - $7.50 thousand. 5,000 up - $7.00 thousand. In cups: 50's - $8.00 thousand; 100's - $7.00 thousand. Add 50c per package postal handling charge. Shipped in packages 1,000, 2,000, 3,000, 4,000, 5,000. Postpaid, prompt shipment, extra count, live delivery, satisfaction absolutely guaranteed. Willow Dale Worm Farm, Rt. 1, Booneville 1, MS 38829.

WHITE Larva, (Spikes), 500 - $2.50; 1,000 - $4.50. **Waxworms,** Webless 250 - $3.50; 1,000 - $11.00. Prompt Postpaid Shipments. **Canadian Nightcrawlers** available. Dealer's Inquire. **Bass Boss Bait Co.,** 1155 Tulip St., Akron, OH 44301.

VIGOROUS Redworms. March special-pit run $5.25 per lb. Super bait 1,000-$6.75. "Happy Hookers" cupped as low as 41c per. Monthly specials. Generous overcount, postpaid. Count on us. H & M Worm Ranch, Box 3623, Simi Valley, CA 93063.

HYBRID Redworms. 5,000 - $15.00; 20,000 - $30.00, postpaid with raising instructions. Brazos Worm Farms, Box 4185, Waco, TX 76705.

VERY Large Redworms. 500 - $4.75; 1,000 - $6.50; 5000 - $29.00; 10,000 - $53.50; small bait redworms: 1000 - $4.95; 5000 - $22.95. Satisfaction guaranteed, postpaid, overcount, Munn, Rte. 3, Box 210, Springfield, OR 97477.

TRAP Nightcrawlers. Earthworms By Thousands. Easy. Instructions. Drawing $1.00. Oldtimer, 117-A Dahlia, Klamath Falls, OR 97601.

GRAY NIGHTCRAWLERS

GRAY Nightcrawlers the Action Worm. Has more moves than a hula dancer. Tough, hardy, resist heat. **They will sell on sight.** We ship the year around. Gray Nightcrawlers breeder or fishing size 1,000—$17.00; 5,000—$77.50; 10,000—$150.00 postpaid. Live delivery. Satisfaction Guaranteed. Free Literature. Telephone (912-835-2542).

FAIN'S BAIT FARM
EDISON, GEORGIA 31746.

FISHWORM Culture — This ... "Raising Worms For Pleasure or Profit" tells how to fix beds indoors, outdoors, small or commercial scale, what, when, how to feed: how and where to sell worms, $1.00 postpaid. Worms, crickets for Sale. Tennessee Worm Hatchery, Section 1, Box 265, Nashville, TN 37202.

MINNOW and Catfish Graders-Agitators - Dip nets - Sienes - Transport tanks. Crescent Manufacturing, Box 3303, Fort Worth, TX 76105.

FREE Picture Folder "How to Make $3,000 Yearly, Sparetime, Raising Earthworms!" Backyard, Garage, Basement. Oakhaven-75, Cedar Hill, TX 75104.

SUCCESS for You! Raise Fishworms and Crickets. Write for Free Literature and our Success Story in Life Magazine. Let us teach you our Secrets of Success. Carter Worm Farm —B. Plains, GA 31780.

NIGHTCRAWLER. Redworm raising. Easy, profitable. New soilless method. Write Charlie Morgan, Box 116-A, Bushnell, FL 33513.

CARTER'S Pure Bred Hybrid Red Wigglers. 20 years of Breeding experience. Top quality. Full count plus extras guaranteed! Write for **Free Literature** on Red Wigglers. Crickets. Mealworms, and 12 inch African Fishworms plus our Success Story in Life Magazine. Carter Worms, Plains, GA 31780.

NIGHTCRAWLERS, Canadian Shipping Via Air, Parcel Post. Thousand Island Bait Store, Alexandria Bay, NY 13607. (315) 482-9903.

LARGE - "Action-Packed" Red Wigglers. Guaranteed live delivery -extras. Special discounts for large quantity. Write for prices-raising instructions, bulk or cup-packed. C&D Bait Farm, Route #2, Box 213-JO, Eatonton, GA 31024.

EXTRA Large redworms - 1,000-$6.00; 10,000-$55.00. Guaranteed Generous overcount. Free instructions included. Postpaid. Canadian-Nitecrawlers available. Bite Best Bait Co., Box 1663, Altadena, CA 91001.

SELECT Redworms 1,000 - $5.95, 10,000 - $46.00. McLean's, 1310 10th Avenue, Arcadia, CA 91006.

ORDER Early, Continuous Supply of Lively, Fat, **Redworms.** $4.50 thousand! Five thousand minimum!! Norb's Bait Farm, 1370 2nd Street, Norco, CA 91760. (714) 737-3459.

WAX Worms, Bee Moths Won't Spin or Cocoon. Color Creamy White. Best Panfish Bait We've Seen. Summer or Winter Price 275 — $3.50, 1,000 — $11.00. Northern Bait, Chetek, WI 54728.

BUSS Bed-Ding Feeds — Keeps — Hauls Worms alive. Wonderful. Free catalog. Buss Manufacturing Company, Lanark, IL 61046.

REDWORMS - Large 1,000 - $6.00; 5,000 - $25.00. Guaranteed Live Delivery, postpaid. Free Literature. Hanus' Wormy Acres, Hayward, WI 54843.

REDWORMS 100 - $2.00; 1000 - $7.00. Breeders $10.00. Postpaid. Godfrey, 1413A Eighteenth, Boise, ID 83702.

EXTRA Large Redworms. 500 - $4.50. 1,000 - $6.50. 5,000 - $27.50. Limited Bedrun Special: 1,000 - $4.25; 5,000 - $11.95; 10,000 - $21.95; 20,000 - $39.95; 50,000 - $89.95. Postpaid, Guaranteed. Bud Kinney, Rt 1, Box 438-T, Chico, CA 95926.

51 FROGS & FISH

YOUR PUBLICATION ADVERTISING

FOR bait and breeding stock sales the classified columns of the outdoor sports magazines have started many an earthworm breeder on a successful career and have been influential in maintaining and increasing his volume from year to year.

They carry a good many competitive earthworm ads, but the readership is large, the market good and the cost relatively small. The publications listed below are the best of that group; the ones which consistently carry the largest volume of earthworm advertising. There are several others, however, sectional and Canadian, which may well be considered by growers located in the areas where their circulation is influential.

BEST OUTDOOR-SPORTS MEDIA

Publication	Address
Field & Stream	2 Park Avenue New York, NY 10016
Outdoor Life	2 Park Avenue New York, NY 10016
Sports Afield	250 W. 55th Street New York, NY 10019
Fur-Fish-Game	2878 E. Main Street Columbus, OH 43209

Information, including closing dates, rates, and circulation, may be obtained by writing the above magazines at addresses given.

By watching the classified columns of these magazines from issue to issue, under the heading LIVE BAIT, you can gain a fairly good idea of what and how your fellow earthworm growers are advertising. A typical column is shown on a previous page.

MAIL ORDER AND SPECIALTY PUBLICATIONS

THE following is a group of publications which reach millions of readers who are generally responsive to classified mail order advertising pertaining to hobbies, sports and home money-making ideas.

They do reasonably well on earthworm advertising, but not as well as the outdoor sports group. They are expensive, because of their big circulations. Try them out cautiously, with test copy, before committing yourself to large expenditures.

Publication	Address
Popular Mechanics	224 West 57th St. New York, NY 10019
Popular Science	1000 Town Center, Suite 1830 Southfield, MI 48075
Organic Gardening	33 E. Minor Street Emmaus, PA 18098
Old Farmer's Almanac	Dublin, NH 03444

Information, including closing dates, rates, and circulation may be obtained by writing the above magazines at addresses given.

Please bear in mind that these large circulation monthly magazines close their forms for classified advertising as much as three months in advance of the date of issue, so you must plan your schedules well ahead. Write for classified rate card to each publication you think you might want to use. Rates change from time to time. Your remittance must accompany your order for classified advertising in all publications.

GARDENING, FARM AND RABBIT PUBLICATIONS

FOR the earthworm grower who is specializing in earthworms for soil improvement, the horticultural publications would seem to be the logical advertising media. . . the gardening and flower growing magazines, and possibly the farm papers. We hesitate to recommend a list of specific publications, however, for the simple reason that advertising returns from such publications usually seem to fall below the level of expectation, and sometimes below the profit level. Advertising that fails to show a profit, or which merely "breaks even", is pointless.

Some growers have done fairly well with sectional farm papers, such as the "Progressive Farmer" and "Farm and Ranch", in the southern states. Other publications may do as well. Our only suggestion is to try them out cautiously, with test ads, and let the results determine your further course of action.

There is one publication we can recommend however, and that is "Organic Gardening", Emmaus, Pennsylvania. Organic gardeners are "earthworm-conscious", and this publication runs frequent articles on the value of earthworms in horticulture. As a result, the returns are usually of better-than-average quality for earthworm advertisers.

The circulation of the rabbit magazines is comparatively small, but the classified advertising rates are low, and for the amount expended the results are excellent. Rabbit growers, in recent years, have gone into the earthworm business on a constantly growing scale, and are good prospects for breeding stock. One magazine that caters to the small stock farmer, including the rabbit raiser, is "Countryside". You may write them for information at 2601 Winter Sports Rd., Withee, WI 54498-9317.

HOW DO YOU WRITE A GOOD CLASSIFIED AD?

THE author has been writing advertising copy, some of it phenomenally successful and some not so phenomenal, for more than forty years, but he still has no sure-fire formula for writing a failure-proof classified ad. Nor, we believe, has any other experienced advertising man.

You may write and rewrite an ad, and refine it until it is letter perfect, only to have it fall on its face. Or again you may dash off a few hasty words, just as they come to mind, because you have to get it off on the two o'clock mail to meet a deadline, and the darned thing pulls its head off!

There are, of course, a few good rules to follow. Make every word count because words cost money, but don't, in the interests of economy, make it so brief that it fails to state your message clearly and invitingly. Write it in such a way that if YOU were the reader you'd want to answer it! On the other hand, don't try to tell the whole story in your ad. Leave something to the reader's imagination. . . something untold for which he will want the answer.

We're not going to try to tell you how to write a good earthworm ad, or even give you examples of how we think they should be written. We might make them sound good, but we couldn't be sure how they would pull for you. We suggest you watch the ads of your fellow growers from month to month. If you see some that appear, without change, time after time, you may reasonably conclude that those ads are getting satisfactory results; otherwise the advertisers would not continue to spend money on them. Don't try to copy those ads, but study them and try to discover what makes them tick.

The most successful ad is not always the one that pulls the most mail. You can offer "foot-long" earthworms and get a bushel basketful of mail every day. It will cost you a fortune just to answer it; but if you can't or won't produce what you advertised, you will have succeeded only in antagonizing your prospects and wasting your advertising dollars. Make your advertising honest and reasonable. . . copy you can back up in your follow-up.

Should you quote prices in your advertising, or simply try for an inquiry? That's another question for which nobody has an infallible answer. Growers do it both ways. If you quote figures which are higher than others quoted in the same publication you may lose some inquiries to the low bidders. Or if your figures are too far below those of other advertisers, you may create the impression that your product is inferior, and again lose some inquiries.

The safer ground, perhaps, is to try for the inquiry. But be sure, before you do so, that you have a good sound, hard-hitting follow-up with which to answer it. It costs money to answer inquiries, and your literature MUST be good enough to turn a fair percentage of those inquiries into orders. Remember that your prospect, when he answered your ad, probably answered several others also, so the real battle for his business is in the follow-up. . . not only price information (the low price doesn't always get the order) but sensible, honest and clean good salesmanship on paper; salesmanship on which you can follow through when you ship his order.

It's a good thing to TEST an ad, by trying it out in one or two publications, before you "shoot the works". If you hit upon one that clicks, keep using it. If it pulls once it will probably continue to pull, even though it is repeated continously in the same publications. A reader may scan your ad a dozen times before it gets under his skin. That's why continuous, rather then spotty or haphazard, advertising pays off. But test changes of copy from time to time. No ad is ever so good that there is not a chance that it might be improved.

HOW TO SELECT ADVERTISING MEDIA

One good rule for selecting media for your earthworm advertising is the time-honored rule for selecting a good place to eat; "look for the restaurant that is crowded!" Look for the publications that continuously carry the most earthworm advertising. Earthworm advertisers have been testing all kinds of publications for years and it's a pretty safe bet that the ones they continue to use regularly are the ones that are yielding the best return. If you feel venturesome, try a new type of publication now and then. You just might hit a good outlet that everyone has overlooked.

Don't cross a publication off your list because it fails to pull on the first insertion. It takes at least three consecutive insertions, possibly more, to constitute a fair test, if the publication shows any promise at all.

HOW TO CHECK AND EVALUATE YOUR ADVERTISING RETURNS

KEYING your classified advertising, and keeping a record of returns from each publication used, will enable you to determine which publications are paying their way and which ones should be dropped from the schedule, after a fair test. Check not only the number of inquiries received (which is not always the best guide to a publication's value) but also record all sales from each ad, using a simple card system.

There are several ways of "keying" an ad. If you have a post office or route box number you can use it as a key by adding a different letter for each publication, as 358-A, 358-B, etc. Or you can use a letter or

Name of Publication ______

Beginning JANUARY 19___ Ending JUNE 19___

Date of Issue	JAN.
Space	20 WORDS
Classification	LIVE BAIT
No. of Insertions	1
Rate	4.50 PER WORD
Cost	90.00
Key	Box 358-B
Paste Copy of Ad Here	

Inquiries	Cash Rec'd	Inquiries	Cash Rec'd
卌 卌	$ 50.00	卌 卌	$ 15.75
卌 卌	30.50	卌 卌	27.00
卌 卌	25.00	卌 卌	19.25
卌 卌	21.50	卌 卌	47.00
卌 卌	57.00	卌 卌	22.00
卌 卌	93.00	卌 卌	67.00
卌 卌	41.50	卌 卌	
卌 卌			
卌			
Totals	$	145	$ 516.50

Cost of space and answering inquiries—$148.00 (29 % of sales)

(The above record is for illustration only - does not represent the actual returns from any specific ad or publication).

number as an affix to your street number. . . as 323-A W. Main St., 323-B West Main Street, etc. It won't interfere with the delivery of the mail, but will tell you which publication drew the inquiry. You may even use variations in your signature, such as Jones Earthworm Hatchery, Jones Worm Farm, Jones Bait Farm, Jones Earthworm Farm.

Keeping a record of inquiries and orders from each ad and publication is good business. From it you can determine which ads and publications are doing the best job for you. It enables you to refine your ads and the publication list to a point where you are getting the best return for your advertising dollar.

Let us assume, by way of illustration, that you place classified ads in 10 national or sectional publications with 9,000,000 total circulation at a cost of $400.00 per issue for the entire group. Let us assume further, that from one month's advertising you obtain only one inquiry for every 9,000 circulation, or a total of 1,000 inquiries at a cost of 40 cents per inquiry. Assume, also, that it costs you 30 cents in literature, envelope and postage to answer each inquiry, making your total cost per inquiry 70 cents, or a total of cost for that month's advertising and follow-up of $700.00. A typical record sheet is shown here.

Let us assume that you'll get at least one earthworm order from each 20 inquiries, a 5 percent return, or a total of 50 orders at an advertising and follow-up cost of $14.00 per order. If those orders average $50.00 (or if you get 100 orders at an average of $25.00), you may feel that your advertising and selling cost has been reasonable and that your advertising has paid off. Your selling cost has been 28 percent, which is about average for most types of mail order selling.

Even if your cost is higher it may prove to be quite profitable over the long pull, due to the "cumulative" value of continuous advertising . . . the prospects who may never answer your ads directly but who become conscious of your hatchery through seeing your ads repeatedly, and who eventually may contact you when they are ready to buy.

Still another dividend from your advertising lies in the fact that initial orders may be followed (and frequently are) by repeat orders from the same customers. In other words, you are building a customer list which can be valuable long after you have forgotten what ad or publication brought the initial inquiry. One GOOD customer won through advertising may be worth far more than the cost of a year's campaign.

One thing more. . . do not be too quick to "close the record" on any classified ad. Some magazines continue to pull for several months; even an occasional inquiry for two or three years or more, though for all practical purposes you may consider an ad "dead" after three or four months.

ADVERTISING AND SALES PROMOTION HELPS FOR EARTHWORM GROWERS

MOST earthworm growers, as they are first to admit, are not skilled in the highly professional requirements of writing or producing advertising and sales literature. . . copy writing, art supervision, type arrangements and the buying of printing or lithography. They need advertising and sales literature in limited amounts, and a small run of any piece of printed matter is costly, as compared to mass production. Original art work or photography, and printing plates, are also costly, when added to the cost of a short run. They are often omitted entirely in the interests of economy and to the detriment of the job.

To meet the need for good illustrated literature at reasonable cost, the author of this book has prepared certain units of literature to meet the most pressing requirements of earthworm advertisers. They are lithographed in large quantites to gain the advantage of mass production prices, and offered in small lots to individual advertisers at far lower cost than if the advertiser were to invest in a comparable job for his own exclusive use.

FOUR PAGE DUAL-PURPOSE FOLDER

This 3,000 word folder, 11 x 17 in; folded once to letterhead size, is designed as a give-away piece; to be used as an enclosure in answering inquiries, or as an instruction sheet to go out with earthworm shipments. For either use it is bound to save you a lot of letter writing, or the costly alternative of preparing and printing your own special folders. It has a generous imprint space for stamping in your name and address, or it will be imprinted for you, at an extra charge, in lots of 1,000 or more. Ask for sample, from Shields Publications, P.O. Box 669, Eagle River, WI 54521.

TO HELP YOU SELL EARTHWORM BOOKS

This brochure describes the complete list of earthworm books offered by Shields Publications.

It is an ideal enclosure to use when answering inquiries from prospective earthworm buyers. Many worm growers find that book sales lead to more worm sales.

Here are the Best of the EARTHWORM BOOKS

The books listed in this revised folder include every book of importance ever written on the subject by such authorities as Shields, Morgan, Holwager, Edwards, Myers and others.

These are books every earthworm grower should have in his "Earthworm Library." Each is by a specialist in some phase of the industry; soil improvement, bait culture, biology and breeding habits, feeding and fattening, advertising and selling... all designed to:

1. HELP YOU GET A BETTER START WITH EARTHWORMS
2. RAISE BETTER EARTHWORMS
3. DO A BETTER SELLING JOB
4. REALIZE A BETTER PROFIT

Each book is moderately priced and will be mailed direct to you on receipt of price.

Send your remittance (check, money order or currency) to:

YOUR BUSINESS NAME
AND ADDRESS

A GREAT BUSINESS BUILDER

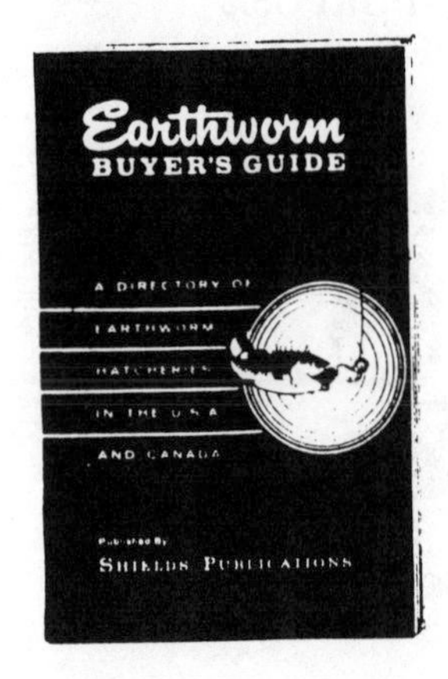

Many worm growers tell us the EARTHWORM BUYER'S GUIDE is their single best investment toward building their business. It is the only directory of worm farms in the U.S.A. and Canada. Each farm listing gives all pertinent information: Name of owner, address, phone number, types of worms or product sold, etc. In addition to being listed in the Buyer's Guide, you can present a more detailed sales story with a display ad.

It is one of the best ways for you to reach wholesale and retail earthworm buyers . . . fishermen, bait dealers, buyers of breeding stock, wholesale bulk worms for stocking beds, or worms for horticultural uses.

The cost of a listing and/or ad is very nominal, and since the BUYER'S GUIDE is purchased only by those who want to buy worms or related supplies, the resulting business it generates can provide an excellent return.

EARTHWORM GROWERS: For cost of the book folder or details about a listing or display ad in the Earthworm Buyer's Guide, write Shields Publications, P.O. Box 669, Eagle River, WI 54521

Some Original Art Work You Are Welcome To Use

SO MANY earthworm growers have asked us for permission to reproduce, in their own literature, various copyrighted illustrations from the original edition of this book that we have decided to make them available to any who wish to use them.

We cannot furnish cuts, because of the great variety of sizes that would be required. We will, however, send a good clear print of any illustration reproduced in these two pages for just $2.00, to cover the cost of print and mailing. Send your order to Shields Publications, P.O. Box 669, Eagle River, WI 54521.

No. 25

PLEASE ORDER BY NUMBER AND SEND $2.00 FOR EACH PRINT YOU ORDER.

SOME ORIGINAL ARTWORK

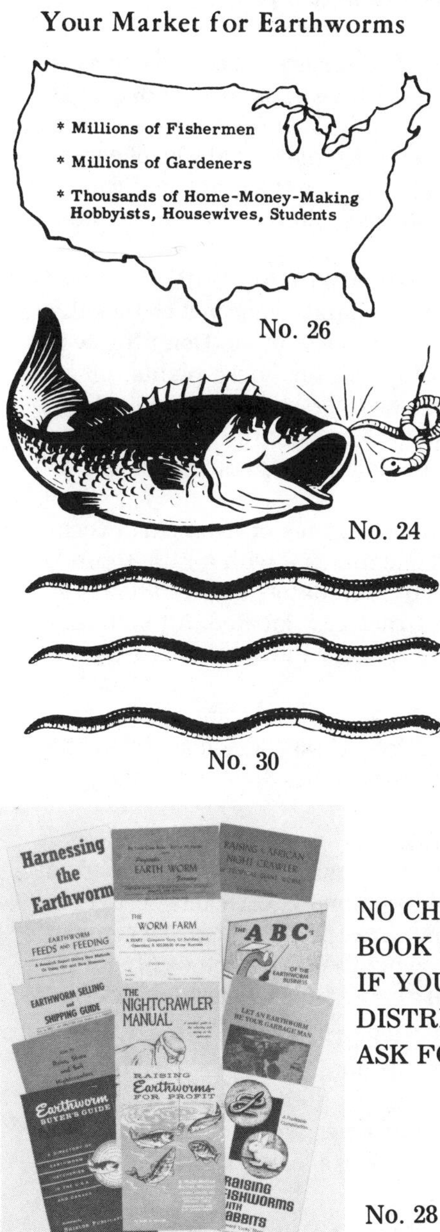

No. 26

No. 24

No. 30

No. 28

No. 29

No. 31

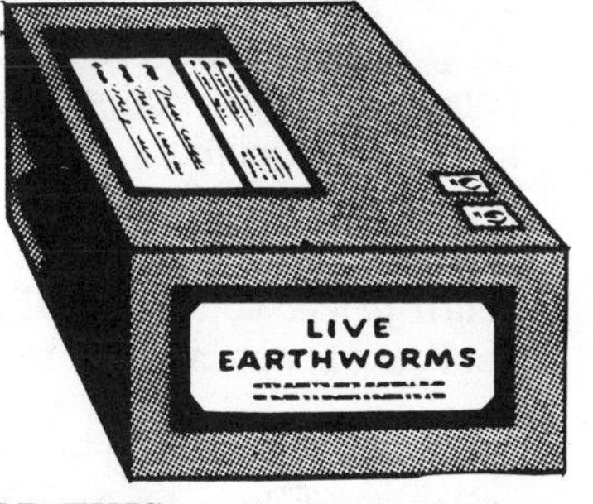

No. 33

NO CHARGE FOR THIS BOOK GROUP PHOTO IF YOU ARE A BOOK DISTRIBUTOR. JUST ASK FOR IT.

No. 27

Supplementary Income from Your Earthworm Project

THERE are a number of products and services closely associated with earthworm growing that may prove a source of substantial supplementary income, with little or no additional operating expense for the grower. Some of these tie in closely with the actual sale of earthworms. Others may provide extra earnings in the slack off-season intervals when earthworm income is normally low . . . the midwinter months.

EARTHWORM CASTINGS FOR POTTING SOIL

This is one product you will have in abundance when you change the bedding in your pits, as you will every six months or so. Don't throw it away! It is the finest imaginable potting soil for house plants, or an excellent mulch for use around flower beds or shrubs. This casting-rich compost may be bagged and sold at substantial prices, either direct to users or through greenhouses, flower shops, seed dealers or garden supply stores.

One nurseryman of whom we know buys tons of used earthworm compost, or bedding, at $12.00 per ton. He mixes it with a little topsoil, a little sand, adds a bit of plant-growth stimulator, and retails it for 65¢ per gallon, bagged in plastic and attractively labeled. A ton makes about 500 bags; retails for $325, a tidy profit for the amount of expense and labor involved.

If you care to advertise and sell your casting-rich compost at retail, or bag it and sell it through retail outlets at 40% discount from the retail price, you will realize a satisfactory extra dividend from your earthworm operation.

OTHER BAITS

If you maintain a retail bait shop or roadside stand, you may find it profitable to add other live baits . . . minnows, crickets, crayfish, grubs, mealworms, etc . . . or you may even arrange to grow some of these. If you have no facilities for growing them you can easily locate growers who will be glad to supply you at favorable wholesale prices.

TACKLE AND FISHING EQUIPMENT

That, of course, is a business in itself; but if you have a local retail earthworm business, you may find it profitable to carry a supply of hooks, lines, leaders, sinkers, floats, cane poles and items of that kind, which some of your earthworm customers are sure to need on almost any fishing excursion.

EARTHWORM GROWING SUPPLIES

In selling breeding stock to beginning growers you will often find a demand for such things as bedding materials, peat moss, worm feed, bait cups and other supplies which the beginners have not yet learned to provide or buy for themselves. There is not a great deal of profit in most of these items; supplying them is more for accommodation than for profit.

They are sometimes an important factor in starting a new grower, however, and if you are in a position to buy in quantity they will pay a modest profit. Feeds and beddings can be sold only locally to advantage; shipping charges are too costly for long range selling.

You may offer, for a reasonable fee, to set up a complete hatchery operation for any beginner who buys his foundation stock from you, if he is within a reasonable distance from you. For such a customer, you would build pits or bins with materials supplied by him, supply the bedding, supply and install the breeding or pit-run worms, water and feed them, and in general get him off to a good start, at the same time giving him a "liberal education" in handling and caring for the worms.

FORMULAS AND INSTRUCTION BOOKS

If, through the years, you have developed a real good feeding formula, make it available to other growers at a reasonable fee but, we beseech you, don't turn it into a "racket"! We've seen so-called "formulas" offered at $100.00 and up, containing the same data that might be obtained inexpensively through any of the good earthworm manuals for $8.00 or less. However, if you have a really good feeding formula, one for which any breeder may buy the ingredients locally, it is at least worth a ten dollar bill. It may take you only a few hours to put it on paper and a few pennies to print it, but that is not what the purchaser is buying; he is buying your years of experience and experimentation; and your formula may earn its cost for him, many times over, within the first season.

If you are sufficiently gifted in the writer's art to author and publish a complete book, covering all phases (or special phases) of earthworm growing and marketing, you will find it profitable to do so . . . and an excellent bit of promotional material. Not all such books are good. We've seen many that told too little and cost too much. However, there have been several books that are well done and really pay off.

Where to Get Supplies for Your Earthworm Growing Project

MANY beginning earthworm breeders are a bit puzzled, and understandably so, as to where they may obtain essential materials and supplies . . . for making worm bedding, for packing and shipping, for advertising and selling, and such items as labels, letterheads, price lists or other printed forms.

Bedding materials . . . manure, peat moss and supplementary feeds . . . are readily available in most communities. You can usually arrange to get cow, rabbit or sheep manure from some nearby farmer, dairy, rabbitry or stock farm and possibly arrange to have a few loads hauled in to you at modest cost. If you know of no such source you can safely use the bagged varieties sold by feed and fertilizer stores. Be sure they have not been chemically treated.

Supplementary feeds, such as poultry mash, calf meal, corn meal, cotton seed meal, alfalfa meal or dairy feeds, can be obtained at any feed store. Elevator and feed mill sweepings may sometimes be obtained at low cost; also bag cleanings.

Peat moss, by the bale or in smaller units, can usually be obtained through garden supply stores. Most nurseries carry it. If your dealers do not carry it in stock, they can order it for you; or you can order it through seed house or nursery mail order catalogs.

As a substitute for peat moss you may, if you choose, use aged sawdust (hardwood, not resinous), ground corn cobs, cotton seed hulls or other waste products — at lower cost, if they are available locally. They are not quite as easy to handle as peat moss but a good substitute.

As containers for packaging earthworms, many breeders use waxed ice cream or cottage cheese cartons, cylindrical in shape, with slip-over lids that fit snugly. Lids can be perforated with an ice pick to admit air. These come in half-pint, pint, quart, half-gallon and gallon sizes. They may be obtained, in large quantities, from a wholesale paper house, or in smaller quantities through a local dairy company or ice cream manufacturer.

In recent years several manufacturers have begun making containers specifically designed for earthworms, attractively designed, illustrated, and with lids already perforated, ready to use. They cost a little more, naturally, than the plain cartons, but they are so much more convenient and attractive, and eliminate so much work and expense in labeling, they are doubtless an economy in the long run. You will note references to some sources for such containers in the chapter on packaging.

Corrugated cartons for shipping may be ordered from a paper box manufacturer or jobber. Or if your needs are limited in the beginning, we suggest that you gather up the shipping cartons that are thrown out by druggists, hardware stores, grocers and other merchants in your vicinity. As a rule they may be had for the asking and would otherwise have to be hauled away and burned. (Corrugated boxes make good worm bedding, too, if you have the facilities for grinding them. Or you can reduce them to pulp by soaking.)

For letterheads, invoices, cartons and shipping labels, sales folders and other printed forms, the best bet is your local printer who may be able and willing to assist you with attractive layouts and arrangements. If he cannot personally do so, he may have contacts with artists, advertising men, photographers or others who can supply professional services when they are required.

No stock materials are available on items such as these, because individual requirements vary too widely to make standardized forms practical or desirable.

This may be a good place to advise our readers that we are not in a position to prepare advertising campaigns or advertising materials for individual earthworm growers, as much as we would like to be able to do so. The pressure of our own work, and the lack of time for special assignments, make it impossible. We have many requests for such services, and deeply regret our inability to handle them.

Nor do we sell earthworms. Because so many growers are valued customers of ours for books and for advertising and sales promotion materials, we cannot in fairness compete with them for earthworm sales. We get hundreds of inquiries and orders for breeding stock, particularly, which we have to return or pass along to other breeders. We're always happy to advise buyers as to dependable sources, of course, but the best advice we know is to obtain a copy of the "Earthworm Buyer's Guide" which lists many good sources, any of which will be glad to send you their prices and complete information.

Fancy Packaging Commands Fancy Prices

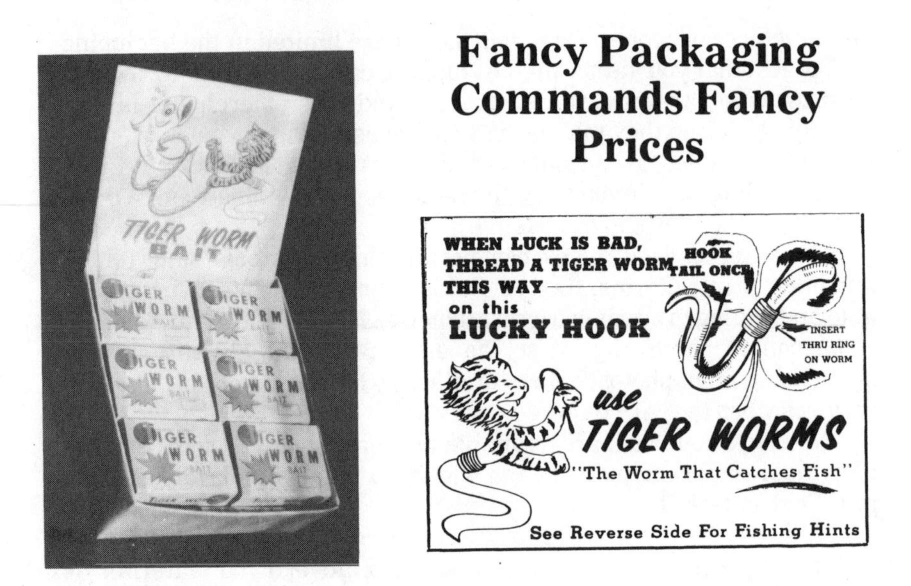

A WESTERN earthworm grower in the trout fishing country hit a new high in merchandising techniques for bait worms and with considerable success. He packs 24 worms in a plastic box with hinged lid, a comfortable and convenient size for a fisherman's pocket — 12 boxes to an attractive counter display carton.

Fitted into the lid of each individual package is a card suggesting the proper way to hook a worm on a size 10, 12 or 14 hook, plus an actual No. 12 hook for the fisherman's use . . . and some fishing hints on the back of the card.

The owner tells us it has doubled their business and is based on the idea of educating fishermen to the use of a small worm on a small hook for best results; an idea we have long advocated in all of our books and manuals, and one we ourselves have found most successful.

The worms so packaged and cartoned are sold through a large variety of retail outlets: tackle shops, boat docks, grocery stores, restaurants, gas stations, bars and some large retail establishments such as Sears .

This kind of packaging and selling is not for everyone, of course. But it deomonstrates the fact that in earthworm selling, as in the merchandising of any other product, originality and ingenuity can be made to pay off. Perhaps it will nudge your own train of thought into new channels, in connection with your own advertising and sales promotion plans.

Earthworm Vending Machines

24-HOUR PER DAY BAIT SALESMEN

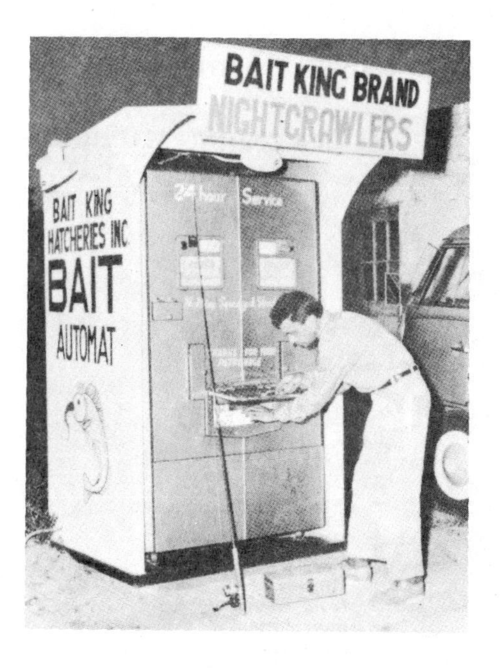

COIN MACHINES will sell you almost anything under the sun, from a Coke or a candy bar to a piece of pie or a bottle of milk . . . so why not a package of bait worms? As a matter of fact, they will!

The idea has been kicking around for quite a while among worm growers, bait dealers and coin machine manufacturers, but only now has it reached the stage of practical usability.

The vending machine has been perfected to a point that promises smooth, trouble-free operation and sales tests are encouraging. A machine in an area west of Chicago is said to have recorded weekend sales of $400.00, mostly in nightcrawlers, a figure which might be cut in half and still look good to any bait dealer or worm grower.

The larger growers, who are in a position to finance the use of vending machines, will readily visualize their possibilities at fishing docks, filling stations or at any other access point to good fishing where an electrical hook-up is available. No attendant is needed, except to re-stock the machine at intervals. A percentage arrangement with the partner who provides the site for the machine and the electrical facilities for connecting it, would doubtless be less than the wholesale discount on the worms if he were to buy them for resale, since there is no labor involved on his part.

Bait dealers may wish to purchase a vending machine outright to serve customers who need bait during hours when the bait shop is closed. This may often be the case since the best fishing is generally early in the morning or late in the day.

The vending machine illustrated here, which may be used either indoors or outdoors, will dispense as many as ten different baits or other fishing items, each at its own price. This versatility broadens the usefulness of the machine. At the time of printing this edition, the list price of the machine is $1,875.00.

A VENDING MACHINE WITH EYE APPEAL

Distributed by The Bait Man, 153 East Woodlawn Avenue, Hinckley, IL 60520, from whom detailed information may be obtained.

This machine is completely automatic in operation and holds 150 cups of any size which can be loaded in a few minutes, according to the distributor. The cabinet is 6'7'' high, 3' wide and 3' deep, is heavy-duty construction and finished with automotive enamel. The refrigerated cabinet keeps bait fresh and lively and is fully insulated. The coin mechanism can be adjusted by the dealer to take in amounts from 25¢ to $5.00.

VENDING WORMS ON THE "HONOR SYSTEM" PLAN

A Kentucky worm grower who lives on a main road to a good fishing lake has a clever worm vending system of his own, and a very inexpensive one. . . coupled with a great deal of confidence in the innate honesty and integrity of his fellow men, and of fishermen in particular. He tells us they have rarely let him down. The record: four boxes of bait taken and not paid for in a period of three and a half years.

The fisherman merely deposits money in a slotted and padlocked metal box set into the cabinet and helps himself to a carton of worms.

The cabinet is 12 x 25 inches, 12 inches high, with slanted plywood top to shed water. It is lined inside with 1½ inch insulating material to protect the worms from the heat of the sun, and is provided with air vents. The worms keep well in 100 degree temperatures. In real hot weather, a sponge in a metal holder is moistened once or twice a day.

The sign above the cabinet is lettered on both sides with Scotchlite tape, which makes it an attractive day-and-night invitation to stop and shop for bait. It has proved to be a profitable and labor-free venture, one that may appeal to other growers in similar situations.

Interesting Facts About Red Worms

1 - They are larger than their "ancestors", the red or mature worms (due to special feeding and care) but retain their natural toughness and liveliness, their attractive coloring and their prolific breeding habits.

2 - They stay at home, under normal conditions. . . will not migrate, or crawl extensively, if adequate food and moisture are provided for them.

3 - They are adaptable to widely varying climatic and soil conditions; do well over a wider temperature range than do native garden worms or nightcrawlers; will live and multiply wherever moisture and organic foods are abundant.

4 - They are bi-sexual, having both male and female reproductive organs. Thus each worm may produce egg capsules, but must first mate with another worm.

5 - Each healthy earthworm, under favorable conditions, may produce an egg capsule every seven to ten days. These capsules incubate in 14 to 21 days, each hatching out from two to twenty worms, with an estimated average of four.

6 - The newly hatched worms will mature to breeding age, though not fully grown, in 60 to 90 days, as indicated by the formation of the clitellum, the thick muscular band about one-third the length of the worm from its head.

7 - The domesticated earthworm will continue to grow after it reaches the breeder stage for as much as six months or so before reaching its full size, and may be further fattened for bait by special feeding. The normal length of a fully matured and well fed red worm is about 3 to 3½ inches; under special feeding it may reach a length of 4 inches or more.

8 - Red worms are the finest of bait worms, preferred by experienced anglers for their color, liveliness and their ability to live and remain active for long periods under water.

9 - Red worms have a life span of several years (when protected) estimated by some biologists at as many as 10 to 15 years.

10 - An earthworm will swallow its own weight in soil or compost each 24 hours and, after extracting the food value, will deposit a like weight in castings rich in plant food values.

11 - Red worms are excellent shippers; they live for weeks, if necessary, in damp peat moss.

Co-operatives. . . Good or Bad?

COOPERATIVES, as applied to earthworm growing, have been tried out in many areas, primarily in the West, with varying degree of success. In most instances they have been short-lived, and not too satisfactory from the standpoint of the grower.

The avowed purpose of the cooperative is to set up a marketing organization for its members, advertising for their collective benefit and allocating equitably among the various members all orders received from such advertising and sales promotion activities.

For this service the association receives a percentage on all sales. In most instances the association or cooperative undertakes to establish new growers, supply them with foundation stock, advise and supervise such growers, and in general to act as headquarters and clearing house for all of the activities and requirements of its membership.

It is the opinion of some growers that cooperatives and associations, in some instances at least, are the "brain children" of their promoters who operate for their own benefit and are prone to exploit the small grower. Membership fees are sometimes inordinately high; the member is bound to purchase his breeding stock and his supplies through the organization and his net return on sales is apt to be reduced to a minimum.

A Western grower writes, with reference to a cooperative with which he affiliated, but which is now defunct:

"They must have seen us coming and said: 'Here comes a sucker'. We paid $600.00 for two 4x7 bins, 6 inches deep; paid $6.97 for one gallon of 'vitamins'; $2.62 for one bag of citrus meal and $2.67 for one bag of walnut meal."

No information as to the number of worms in the two 6 inch bins, but the management obviously made a good profit on them; and as to the "extras" included in the deal, most successful growers probably never heard of, or found it necessary to use, these additives for their worm bedding.

A highly regarded veteran earthworm raiser, who has observed cooperatives for many years, writes: "I have seen no benefit from it (the cooperative) except possibly to a few of the larger operators. There are not enough producers in any one locality to justify an organization. The best that can be said for an organization is that it would provide a forum for the exchange of ideas. Earthworm culture is peculiarly an individual business. Each grower must find or create his own market."

A truly cooperative effort, through a non-profit organization developed by the growers themselves rather than by outside interests, could be of great benefit to the member growers, many of whom admittedly have neither the resources nor the business training for building up volume sales for themselves. The difficulty here lies in the fact that it would be hard to find one of the group who might have the skill and the available time to act for the entire membership, even on a paid basis.

We've found that in many instances friendly fellow growers in a community have worked out a satisfactory system of cooperation of their own, without benefit of an organization of any kind. They supply one another with needed stock, on occasion, either at reasonable cost or on a split-profit basis. They talk over their problems, exchange ideas, and in general observe the attitude of "good neighbors" rather than that of "competitors". In this writer's opinion, that's the way the earthworm industry is best conducted. . . for the benefit of all. In the long pull it will pay off, both in good will and sales, even for the most "rugged individualist."

The opinions expressed in this chapter are the writer's own, not intended in any way to discredit the idea of cooperatives, or any such existing organization that may be operating on a truly "cooperative" basis for the benefit of all of its members, and with a fair return to the management for the services performed.

We suggest, however, that before you join such an organization you investigate it thoroughly, talk to other members, analyze its benefits to you as a grower, as compared to doing your own advertising and selling, and have full access to all records. If it is a cooperative, it is, after all, YOUR organization.

Most growers find it advisable to "go it alone"; with the advertising and sales helps now available (not accessible to the small grower a few years ago) it isn't difficult to do just that.

Earthworm Breeding As A Retirement Project

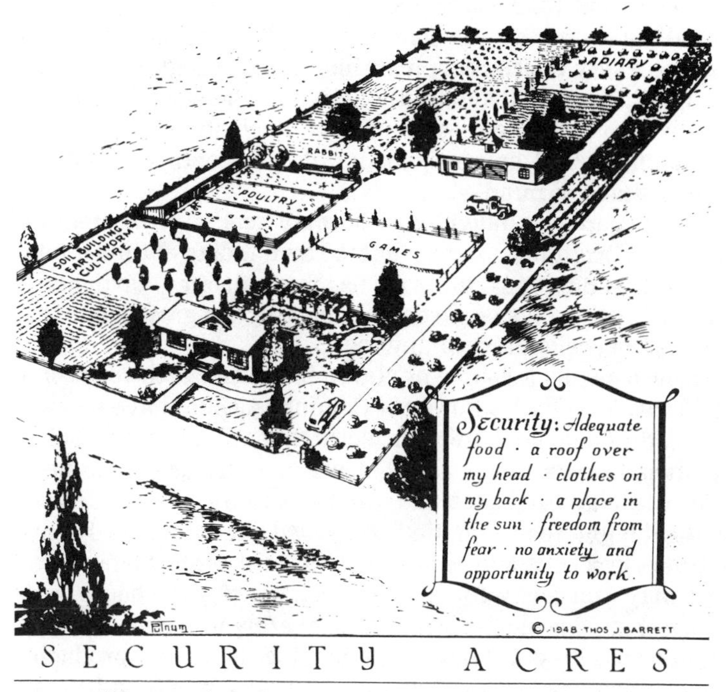

Illustration Courtesy Earthmaster Farms

Illustration Courtesy Earthmaster Farms

"RETIREMENT" means many things to many people: to some, a fateful, discouraging, even a terrifying prospect; to others, a promise of happy years of self-determination, the joys of placid living, the fulfillment of lifelong dreams.

To some it means the enforced cessation of accustomed duties and the loss of accustomed earnings, regardless of the worker's ability or inclination, because of employment rulings. Sometimes it threatens a real hardship.

To others it represents a welcome opportunity to get away from the daily grind of time-clock regularity or the unending routine of desk and swivel chair. . . to do, in the golden autumn years, the things they've always dreamed of doing; an opportunity to live fully, leisurely, comfortably, amid pleasant surroundings and in the congenial fellowship of good friends and neighbors.

Some are privileged to retire with the comforting assurance of pensions or other income adequate to all material needs; others must search out new sources for earnings that will partially or fully meet the week-to-week and year-to-year costs of living.

But all have one thing in common. . . whether financially independent or not, every retired couple or individual recognizes the need for some kind of interesting and useful activity, whether it be entered into as a hobby or as a living-income project. Such stimulating activity or employment, mental or physical, is vital to continued health, happiness and peace of mind.

For the couple or individual who may seek a retirement project that is useful, profitable, and yet not too demanding of time or physical energy, the breeding of earthworms has much to offer, particularly if it is combined with a love of gardening, pleasure in growing things, or a yen to do an occasional bit of angling.

Earthworms, as a couple of clever feminine earthworm breeders (Mary Crowe and Gladys Bowen of Columbus, Ohio) have recounted in their literature, "do not bark, bite, scratch, howl, squeal, quack, crow, cackle, moo, bray or neigh, smell, screech, chirp or call names."

They don't have to be fed and watered twice a day, bedded down or locked up at night, fenced in, tied up, turned out, or generally cared for on the time-clock schedule required in the care of most other livestock.

You may run away for a long weekend if you choose, with the assurance that your earthworms will be safely in their beds when you return, and that they will have been feeding, breeding, producing egg capsules, hatching new earthworms and GROWING, all at a dizzy pace, while you have been away.

Yet it is doubtful whether anything else you might grow, animal or vegetable, could promise a more profitable return on so modest an investment, and upon the limited amount of space, time and labor involved.

There's plenty of work to be done, of course, but it's not too hard, too exacting or too confining. And for the person who loves the soil, who likes to experiment with growing things (or who likes to go fishin'!) the breeding of earthworms can be sheer FUN!

An acre or two, a good garden, and EARTHWORMS. . . these, to many who retire, may spell S-E-C-U-R-I-T-Y.

Questions and Answers

QUESTION: Can I use native worms from my own soil for breeding stock?

ANSWER: Yes, but they adapt themselves slowly to life in captivity, and because they are relatively slow breeders they are not regarded as commercially profitable.

QUESTION: Can I breed native nightcrawlers in captivity?

ANSWER: Native nightcrawlers can be held for long periods in pits and will breed to some extent, but too slowly to make such an operation commercially profitable. Not only are they slow breeders, but their breeding habits differ from those of the red worms in that they lay their capsules in a permanent system of burrows branching off the main burrows. Digging the beds disturbs the system and interferes with normal breeding activities. Lawns or other areas to which the crawlers are native can be specially fed and watered to increase their population.

QUESTION: Can I raise earthworms in my compost heap?

ANSWER: Yes, and with much benefit to the compost and the soil on which it is used, because the earthworms aid materially in the rapid breakdown of the organic matter in the heap, at the same time enriching the compost with their castings. CAUTION: do not introduce the earthworms into the heap until it is past the heating stage.

Also note that this is merely a supplement to, and not a substitute for, the bed or pit method of raising earthworms, because it would hardly be practical to recover the earthworms from the compost heap for packing and shipping. Furthermore, they may escape, to some degree, from the heap.

If you are making compost for earthworm bedding, however, the presence of earthworms would be beneficial, and both earthworms and spawn would eventually find their way into the breeding beds as you use the compost.

QUESTION: Can fresh manure be used in the earthworm beds?

ANSWER: Fresh manure may be used if it is spread over the top of the pit material where there is no danger of heating and where the worms may feed on it from the under side.

QUESTION: Can fresh grass clippings and leaves be used in the earthworm beds?

ANSWER: No, because they may heat. Let them dry first, then mix with soil, or compost them before using. Leaves should be shredded, if possible, to make them more readily available for either earthworm food or composting.

QUESTION: Do earthworms eat or injure living plant roots?

ANSWER: No, they eat only matter that has lived and died. They doubly benefit living root systems, first by tunneling the soil to permit root expansion, and second by enriching the soil with their casts and providing available plant food.

QUESTION: What vegetable or animal wastes, if any, are harmful to earthworms?

ANSWER: Most breeders advise against the use of citrus fruits or skins as being too acidic; also eucalyptus and pepper leaves, and leaves or grasses that have been treated with poison sprays. Avoid high concentrations of vinegar, salt or strong alkalies.

QUESTION: Do earthworms have any specific breeding seasons?

ANSWER: No. They will breed continuously under proper temperature conditions if sufficient food and moisture are available, as in basements, caves, etc., where temperatures range from 40 to 70 degrees. In their natural habitat, breeding slows down or lapses in mid-winter or in extreme summer heat.

QUESTION: Do egg capsules always hatch in the normal 21 - 30 day period?

ANSWER: No. If temperatures are too low or the beds are too dry, the capsules will lie dormant or hatch more slowly until proper conditions are restored. Capsules may, in fact, purposely be held dormant by refrigeration or drying, then hatched at a later date by restoring normal temperature and moisture conditions.

QUESTION: What is the earthworm's normal life span?

ANSWER: We don't believe that anyone knows the exact answer, but in practice breeders are known to live and produce capsules for several years. Some authorities believe that earthworms, under ideal conditions, may live as long as 10 or 15 years.

QUESTION: How many new earthworms will each breeder spawn in one year?

ANSWER: One active breeder, under favorable year-'round food, moisture and temperature conditions, will produce approximately 50 to 60 capsules, which should hatch about 200 young earthworms. But that's only half the story, because the young earthworms themselves become breeders within three to four months from the capsule stage, resulting in a rapid pyramiding of production before a year has passed. Thus a single breeder, with the assistance of its rapidly maturing children and grandchildren, may beget a total of 1,000 to 1,500 progeny (conservatively) within the 12-month span.

QUESTION: Are there any climatic restrictions to the breeding of earthworms?

ANSWER: No. They are raised successfully under proper controls from the Arctic to the tropics. In cold climates the outdoor breeding season is shorter, and they should be protected in winter by heavy mulching. Indoors, where temperature can be maintained at 40 degrees or above, they will breed continuously, regardless of climatic conditions.

QUESTION: How far may earthworms be shipped successfully?

ANSWER: Earthworms are being shipped successfully, by American breeders, to every part of the world. For domestic shipments, parcel post or United Parcel Service are satisfactory. For foreign shipments, air freight is the only safe method. It is costly, and carriage charges should, of course, be borne by the buyer. In all shipments it is important to provide, in the compost or bedding, sufficient moisture and food to see the earthworms through to their destination.

QUESTION: Can earthworms withstand radical changes in environment and feeding?

ANSWER: Mature earthworms will suffer from such abrupt changes, but will adapt themselves gradually. The breeder should advise his customer in the making of bedding material similar to that in which the breeder stock has been reared. The young worms hatched from egg capsules, however, will rapidly adopt the soil and food conditions under which they are hatched.

QUESTION: How much will it cost me to get started with commercial earthworm breeding?

ANSWER: Very little if you are willing to start slowly and conservatively. More about that in the chapter entitled "How To Get Started With Earthworms."

This hot weather worm has gained in popularity as a large bait worm that can be raised commercially.

Profitable Newcomer to the Live Bait Industry

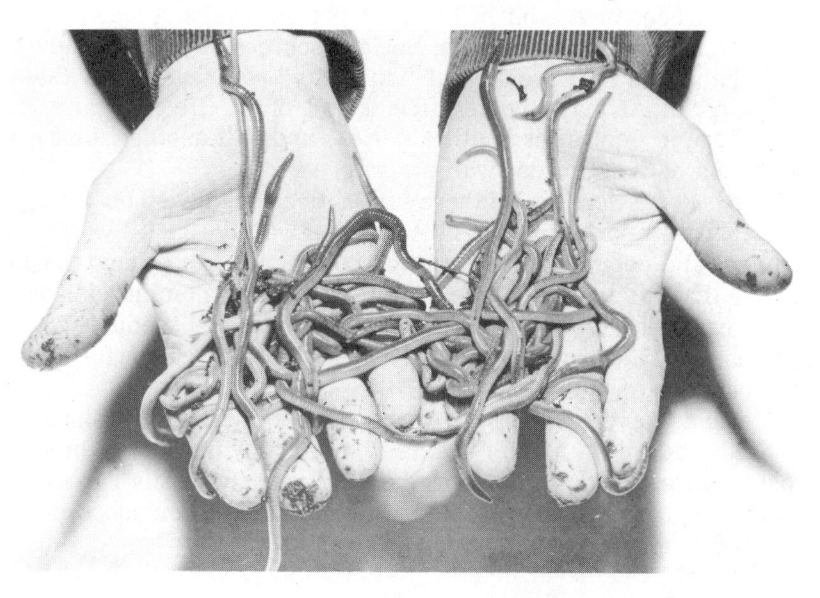

Photograph courtesy L. B. Rand

A COMPARATIVELY new development in the field of commercial earthworm growing, the African Nightcrawler, because of its greater size, its adaptability to pit culture, and the premium prices it commands as a bait worm, has created a tremendous amount of interest within the multi-million dollar live bait industry.

So great is that interest, in fact, that Africans are now being raised, experimentally or on a commercial scale, in almost every part of the United States.

This section is designed to give you, in capsule form, the essential information about this popular big bait worm, and how to grow it.

ONE WORD OF CAUTION: If you are now a red worm grower, don't plan to give up your red worm business completely in favor of Africans. Work them simultaneously, if you choose, and make all the money you can on both. In the final analysis, the good dependable, easy-to-raise red worm is still, and we believe will continue to be for a long time, the backbone of the earthworm industry.

The African Nightcrawler

THE "African's" name is doubtless a misnomer, but one that has become established through popular usage. A maze of conflicting information comes from various sources as to the origin or native habitat of the species. One "authority" states definitely that it originated in Africa; another, the island of St. Helena; another, somewhere in Central or South America; and still another, that it is probably a native Florida swamp worm.

The best answer to all of these suppositions might well be. . . "Who Cares!" The important thing is the fact that we have such a worm; an excellent and popular bait worm that commands exceptional prices, and one that can be raised successfully in pits on a commercial scale (under carefully supervised conditions). Up to now, it seems to be the best answer to the insistent demand on the part of fishermen for a BIGGER bait worm.

Being a tropical worm, the African is more difficult to raise than the hardy red worm, and for that reason probably never will threaten (volume-wise) the supremacy of the abundant and easily raised red varieties, or achieve more than a fraction of the tremendous multimillion dollar annual earthworm sales volume.

It breeds well, nevertheless, grows rapidly, and for those who are equipped to handle it the African offers an interesting and profitable potential earthworm income, or a rewarding addition to the present red worm set-up.

For the simple reason that it IS more difficult to raise than is the red worm, it is safe to assume there will never be an over-production of Africans, or any necessity for making price concessions in order to dispose of all that can be grown.

The African's chief handicap, in commercial production, lies in the fact that it is super-sensitive to cold. It must be raised under carefully controlled temperature conditions, with 50 degrees the lowest safe minimum and with a UNIFORM bedding temperature range of 75 to 80 degrees for maximum production.

That may mean an upper room temperature near the ceiling of at least 90 degrees, because the bedding is always a few degrees cooler, as is also the temperature near the floor. Bedding temperatures should be checked periodically by inserting a thermometer deep into the bedding for a minute or two.

HOUSING THE AFRICAN NIGHTCRAWLER

THE following information applies, primarily, to the cool and cold areas...everywhere, in fact, except the narrow semi-tropical

Corner of a Missouri African Nightcrawler building; insulated, gas heated; equipped with concrete floor pits for growing and fattening, and elevated bins for breeding and hatching. Photograph courtesy of Adams Worm Farm.

band across the extreme southern part of the U.S.A. It goes without saying that the African must be raised indoors, except for the southernmost part of Florida, and even there heavy losses may be experienced from unaccustomed cold waves.

Existing buildings, if well insulated to prevent the undue loss of heat, may be utilized. An uninsulated building is unsuitable because the maintenance of uniform heat in such a building is difficult, and sometimes so costly as to be almost prohibitive. A basement may be used if it is suitable for thermostatic heat control.

If present buildings are not suitable, a building of inexpensive construction may be erected, preferably of masonry materials. . . blocks, tile, concrete or brick. Wood construction is not so good, due to the fact that the combination of heat and moisture causes "sweating" which is bad for wood walls and insulation. If wood is used, however, inside walls should be covered with roofing felt and a moisture-proof lining of some kind to prevent deterioration.

Vented gas heaters, for either natural or LP gas, thermostatically controlled, are probably the most practical as well as the most convenient and economical for African Nightcrawler propagation. Any other type of heat is equally good if it can be controlled day and night, but manual control is likely to be erratic and any lapse of long duration may prove costly.

Draft-free ventilation should be provided, either by means of ventilator windows or by automatic type ventilators atop the building, free from direct drafts. All outside doors to the building should be kept locked to prevent any possibility of their being accidentally left open when the attendant is not on hand to guard them. An automatic door closer, or a strong spring, will serve the same purpose. An open door can quickly chill the room or cause the heating equipment to work continuously — and expensively!

PITS AND BINS FOR AFRICANS

SOME growers advocate masonry type pits (cinder or cement blocks, or poured concrete); others maintain that wood bins are more productive. The worms will do well in either if proper temperatures and moisture conditions can be maintained.

Temperature control is more difficult in masonry pits, however, because they must be built at floor level with no means of circulating warm air beneath them. For that reason they are less efficient for capsule production, for maximum capsule hatch, and for young worm culture. They are suitable, however, for feeding and fattening partially grown worms for bait, and so may be used to advantage in conjunction with wood bins.

Photo courtesy Marvin A. Dickman

In arranging a combination of masonry pits and wood bins the pits are built first on the floor, usually two blocks deep (16 in.) with runways of convenient width between the pits. Keep them away from outside walls which conduct cold.

The wood bins are built directly above the masonry pits, at about eye level, supported by two-inch galvanized, rust-resistant pipe legs, anchored in the corners of the pits and at intervals along the length of the pits to give a sturdy support. The wood bins, previously built, are set on these pipe legs and anchored securely by means of pipe flanges.

The construction of the wood bins may be varied to suit the builder. A suitable construction is a frame of 2x2 inch material, with sides and ends of 1x12 inch lumber and a reasonably solid board bottom.

Bottom and sides should be lined with asbestos board or masonite, or other protective material to prevent direct contact between the wood and the damp compost. Asbestos board seems best; less apt to warp or buckle. It is brittle, however, so be careful in nailing it, or drill it for nail holes.

The advantage of these elevated bins, which may easily be worked from a low bench or from planks across the tops of the pits below, lies in the fact that the air is naturally warmer at the higher level; and that this warm air circulates completely around the bins, making it easy to maintain the even 75 to 80 degree temperatures at which capsule production is at its peak and at which the capsules hatch best.

For small scale production or for an experimental operation, Africans may be grown successfully in tubs, in oil barrels cut in half, wood boxes, or any other type of container that may be available. Whatever container you use, provide for drainage and set them up off the floor to permit warm air to circulate completely around and beneath them.

AT LEFT:
An elaborate three-tier African Nightcrawler installation which multiplies the capacity of the floor space by "bunk bedding" the worms. This system has other advantages also as noted in the text above. Photograph courtesy of African grower L. B. Rand

BEDDINGS AND FEEDS FOR AFRICANS

FOR the busy grower who has little time for gathering special materials for beddings, the simplest and easiest formula is ½ manure and ½ peat moss, mixed, moistened and turned until there is no danger of heating.

Cow manure is probably best and easiest to obtain, though rabbit or sheep manure may be used if available. Peat moss alone, with an abundance of supplementary feed, may also be used for fattening the worms for bait, but the worms will breed best and throw more egg capsules if manure is used liberally.

Dried and crushed leaves (no walnut, pepper or peach leaves, or pine needles) plus dried grasses and lawn clippings, or ground hay or straw, are good additives for the bedding. The worms will also consume the peat moss and reduce it to a rich black humus, but there isn't too much food value in it. It provides bulk (like hay for cattle), but its value is chiefly as a habitat for the worms. It also has the ability to hold moisture and keep the bedding loose.

Because Africans are surface feeders, and because of necessity the pits or bins must be indoors and artificially heated, the bedding need not be deeper than 4 to 6 inches. If the bedding is too deep the air does not penetrate to the bottom of the bin.

African Nightcrawlers are prodigious feeders, so replenish the feed as often as the previous feeding has been consumed. Fishermen like plump worms, so plenty of feed is a good investment. It is better to feed lightly and often than to feed heavily and infrequently. If the worms clean up each feeding completely the likelihood of mold in the pit will be much reduced. A layer of leaves or straw over the bedding, after feeding, will hold moisture and also keep the feeding area dark so the worms will feed more readily. This cover may be raked off before each new feeding and later replaced.

One of the best supplementary feeds for Africans is fresh cow manure, watered down to a fairly thin mixture and poured in strips on top of the bedding. Sounds "messy", and not too fragrant, but the worms grow fat and fast on it, and properly handled it is not quite as messy as it sounds. A semi-liquid mixture of dried sewer sludge may also be used in strips, for top feeding in the growing and fattening pits. It is not recommended for the breeding pits or bins; it may slow down breeding.

The natural feeds in the bedding should be supplemented by the use of ground grain feeds, such as poultry mash, spread thinly over the surface and fed at frequent intervals; daily, as a rule. Water the bedding BEFORE the supplementary feed is applied. By applying the feed dry the worms will consume it from the under side as it becomes moist, and in this manner maggots will be discouraged.

One successful breeder alternates his feeding methods by feeding poultry mash one week and liquid manure the next. The danger in feeding poultry mash continuously lies in the fact that it attracts a troublesome horde of maggots and insect pests.

Check frequently for acidity in the bedding. It is apt to be high where peat moss is used as a base, supplemented by any of the ground grain feeds, and should be counteracted by frequent applications of powdered limestone (calcium carbonate). An excess of limestone has no apparent ill affect on the worms, so there is little danger of using too much of it. Some breeders mix one pound of limestone with each ten pounds of feed, and so apply it with each feeding. You can test for acidity, if you choose, with a soil testing kit or with litmus paper available from your druggist.

As in raising red worms, the bedding for Africans should be turned at regular intervals, just as you would spade a garden. This keeps the bedding loose and fluffy, aerates it, helps to prevent acidity and discourages insect pests that thrive on acid compost.

STOCKING AND WORKING THE PITS

When pits or bins are ready for stocking, and filled with 4 to 6 inches of bedding of proper temperature, introduce young banded breeders at the rate of about 30 or 40 to each square foot of pit surface. Thus a bin 3 x 10 feet should be stocked with approximately 1,000 to 1,200 breeders.

Each bin so started should produce within a year, under favorable conditions, at least 15,000 to 20,000 bait size worms . . . an increase of some 1,200 to 1,500 percent . . . plus enough spawn to stock several additional pits of similar size.

FOR EXAMPLE: A Midwest African breeder who started less than a year ago, as this is written, states: "I sold 450,000 African worms this year from a start of 32,000 breeders with which I stocked eight beds last December. They all hatched out in just a two month period, from January 1st to about March 1st. I've stocked the same beds again with 25,000 breeders this time, during the last few days of October, and now young worms are beginning to hatch out again (middle of November).

"I had some of the beds too crowded with breeders last year, so I think the 25,000 breeders will produce as many young as did the 32,000 last winter. Still have lots of good sized Africans on hand, from the second hatch early this year."

Another breeder started on August 1st with 2,000 breeders, and through December 31st the following year sold 124,000 bait worms with enough left over to restock six 3 x 12 foot beds, a phenomenal 6,200 percent increase in about one and a half years.

From the capsule stage, bait size worms will develop within four to six months, and at that stage they will also begin to throw a tremendous number of new capsules.

Whenever a bin is loaded with egg capsules and young worms, the breeders should be removed and sold for bait or used to stock additional bins. It is well, of course, to limit the sale of bait worms in the beginning in order to stock new pits to the capacity of the building in which they are housed. From that point on you'll have to sell worms fast to keep abreast of your production.

Although Africans may grow to a length of 12 inches or more at maturity, they are usually sold for bait at about half that length, the size at which they are liveliest and best for bait use. Also, it is believed that young worms of this length, or whenever they are banded, are best for breeding stock. The older worms are likely to grow fat and sluggish and diminish in capsule production.

Pits well stocked with worms of varying ages may be worked, and the bait size worms removed, at intervals of about five or six weeks. The number of worms of salable size will decrease with each subsequent working; after three or four workings the pit should be restocked, using fresh bedding.

TEMPERATURE CONTROL

BREEDERS find that at bedding temperatures below 75 degrees, there is a costly mortality among egg capsules, resulting in a poor hatch . . . also a mortality among the delicate newly hatched worms. At the proper temperature, however, the hatch will be nearly 100 percent, and the compost so literally swarming with young worms that one may pick up three or four hundred of them in a single handful of compost.

That's where the elevated upper tier of bins comes into good play, utilizing the warm upper air of the heated building. Some growers have gone a step farther and have built a third tier of bins near the roof. That, however, has proved to be impractical, because at or near the ceiling the air gets TOO HOT. At bedding temperatures above 85 degrees the breeder worms will not feed properly, nor will they breed and produce egg capsules normally. This may be just as well because third tier bins are awkward at best and difficult to work.

This is the procedure in handling the combination of masonry floor pits and elevated bins of wood construction:

Breeders are placed in the elevated bins in 4 to 6 inches of compost, where they remain until the bins are loaded with capsules and hatching young worms. The breeders are then removed to other elevated bins to begin a new cycle.

The young worms remain in their home bins until they are about 2 inches long, then are removed to the well warmed compost of the lower masonry pits for feeding and fattening to bait size. For this purpose the somewhat cooler temperatures are not detrimental, and the twin capacities of floor pits and upper bins, as the cycle is repeated continuously, give the breeder a tremendous volume potential within a relatively small space.

HEATING BEDS WITH CIRCULATING HOT WATER

Ozark Worm Farm in Missouri uses a gas heated 30 gallon hot water heater to circulate hot water through copper tubing or plastic pipe laid near the bottom of each pit or bin, four rows to each bed. It works wonderfully well, they tell us. The one 30 gallon unit heats 16 large beds, with the heater operating about half the time. When set at 110 to 120 degrees it heats the beds uniformly at 75 to 78 degrees, and the water loses only about 10 degrees in making the round trip. An electric circulating pump keeps the water moving. With this arrangement it is unnecessary to keep the room at such high temperatures as were formerly required and probably means less total gas consumption, though we have no figures on that.

RAISING AFRICAN NIGHTCRAWLERS IN THE SEMI-TROPICAL "DEEP SOUTH"

Tree shaded open pits in central Florida. Pits are made of 1 x 8 inch cypress boards. In the background: permanent building for winter operation; a removable wall shed that can be closed and heated. Photo courtesy Justison Worm Farm.

In the southernmost areas of the United States, Africans can be raised successfully in outdoor pits for the greater part of each year, but the southern tip of Florida is the only area where they can be raised in complete safety without some protection against unexpected cold waves.

In addition to the cold hazard, open pit African growers are confronted with the prospect of heavy losses through "crawling", the tendency of the larger worms to migrate on dark nights and in damp weather.

For these reasons, even the growers in the semi-tropical areas are coming to the use of buildings that can be enclosed and heated in winter, as indicated by some of the illustrations shown in these pages.

The biggest market for southern African growers is in the Midwestern and northern states, where there is a lively demand for both bait and breeding stock. Northern growers ship in millions of Africans each year for resale to supplement their own production of Africans and red worms.

LEFT - Removable wall shed for summer and winter breeding and storage. Can be heated in winter. Justison Worm Farm.

BELOW - Work bench for packing Africans; building 30 x 50 feet; has large pit capacity. Equipped with oil stoves for winter heating. Justison photo.

Typical back yard or experimental setup; partially shaded but lets some rainfall come thru. Charlie Morgan.

HARVESTING THE AFRICAN NIGHTCRAWLER

Picking Africans from open pits in central Florida. Justison Worm Farm.

MOST African growers follow the time-honored custom of hand picking and counting the worms, a method not nearly so slow or tedious as one might surmise. Pickers become amazingly adept and speedy with a little experience, and the Africans, being large, are easier to sort and handle than a smaller worm.

Since African beds are shallow and the worms are usually within the top 4 inches of the bedding, many pickers work right from the beds, stirring and turning the compost by hand as they go, as illustrated in the accompanying photograph. That method, however, is pretty hard on the back muscles!

An easier method is to work on benches or tables of convenient height for standing or sitting, using large flat, shallow trays for bringing the worm-filled compost from the pits to the tables.

As each tray is emptied, it is refilled or replaced by another. It may contain several hundred bait or breeder size worms. The average skilled worker can count and package, ready for shipping, about 2,000 worms per hour.

If the worms are to be packaged for retail bait they may be counted directly into the containers in 25s, 50s, etc.; or for bulk shipment, about 350 bait or breeder size worms to each gallon; pit-run, 400 to 500 per gallon, dependent upon size.

PACKAGING AND SHIPPING AFRICANS

AFRICANS may be packaged and shipped in the same manner employed in shipping red worms when the weather is reasonably warm. They may be insured and should in most instances be mailed "Special Handling" when sent by parcel post to expedite delivery. However, United Parcel is preferred by most worm growers.

In the Midwestern and northern states, or in shipping from the deep south into a cold area, the shipping of African Nightcrawlers is a seasonal matter. They can be shipped safely only during the months in which they will not be subjected to temperatures much below 50 degrees.

Actually, the shipping cartons will maintain temperatures, in cool weather shipping, a little higher than those of the outside air; they cool slowly. But if they are addressed to an area where temperatures are apt to be too low for safety, the buyer should be warned not to transfer them to his beds until the bedding and building temperatures are sufficiently warm for them, 70 degrees or more. Many losses occur at this stage of handling.

As in shipping other types of worms, peat moss, pre-soaked and squeezed fairly dry, is standard for African Nightcrawlers. Canadian or German peat is the best. The peat moss should be pre-soaked for at least 24 hours so the fibers will be thoroughly impregnated with water. If warm or hot water is used, the job can be done in a small fraction of that time. It should be squeezed reasonably dry with the hands before packaging the worms.

TEACHING HIS NIBS, THE AFRICAN, TO STAY AT HOME

The African likes to crawl and will attempt to escape if food, moisture or temperature conditions are not to his liking; or for no good reason at all, if he decides to migrate. The main deterrent to crawling is the constant use of lights over the pits at night or on dark or rainy days.

Some growers construct narrow overhanging ledges of wire screen around the top of the pit, just below the rim, to prevent the worms from climbing out. Other devices to accomplish this are: (1) a band of strong yellow laundry soap solution (or sheep dip) painted around the inner rim of the pit, and renewed at intervals (the Africans don't seem to like to cross such a barrier) or (2) a mildly electrified, dry battery charged wire, stretched around each pit just below the inside upper rim. It's the electric livestock fence principle, reduced to worm dimensions. In your own experience you may find the use of lights sufficient as many breeders do. But it is well to know about these other methods.

ADVERTISING AND SELLING AFRICANS

SECOND only to the demand for Africans as bait is the demand for breeding stock, primarily by established earthworm growers who want to raise Africans as an addition to their present red worm line. They recognize the demand, on the part of a considerable segment of the angling fraternity, for a larger bait worm, and they want to be in position to meet that demand when it arises or when it threatens to offer competition.

The first step in establishing your hatchery as a source of supply for either bait or breeding stock is to make sure you are properly listed in the Earthworm Buyer's Guide and Directory, the buyer's guide for the entire industry, and for the hundreds of new breeders who come into the industry each year. It is published by Shields Publications, P.O. Box 669, Eagle River, WI 54521.

While your production is still small you may be able to dispose of all the Africans you can raise, either through local advertising or through personal contact with bait dealers or an attractive sign in front of your hatchery. If you are in the heart of a good fishing area, you may prefer to "run a route" to supply bait dealers personally.

When you are ready to reach out for a wider market, you will want to consider national advertising via the classified columns of the outdoor sports publications, such as Outdoor Life, Field & Stream, Sports Afield and Fur-Fish-Game. They have a combined circulation of some 4,500,000 sports-minded readers, including most of the live bait dealers of the nation, your best prospective customers for Africans.

You can reach additional millions of readers, if you choose to do so, through other national publications, plus some that are local or sectional, but the ones named above will doubtless be your first line of attack.

Classified is more productive than display advertising, dollar for dollar. The classified reader is "looking for something", products or opportunities, and is less likely to miss your ad.

You don't have to be an advertising genius to write a productive ad on Africans. As a rule the words "African Nightcrawlers" or "6 to 7 inch bait worms" are magic in themselves. What else does a fisherman need to know? Quote prices or not, as you may choose, or offer (in national advertising) to mail your price list and other literature. You'll get lots of mail!

GETTING OFF TO A GOOD START

YOU may find it difficult, in the beginning, to build up your African population rapidly enough to take care of the demand, a demand that may amaze you. Your own project, if you start on a modest scale (which is advisable), will require at least a year to get into high gear so you can sell in volume, but there is another profitable way to build up your sales volume right from the start.

Many beginning northern breeders find it advisable, while waiting for their own breeding operations to reach volume proportions, to ship in from the deep south or from established northern breeders enough young Africans (at bulk wholesale prices) to stock or restock their pits.

These worms are usually purchased at the pit-run stage, young worms that can be stored and fed for six weeks or so until they reach bait size. Such worms, delivered, will cost around $9.00 to $10.00 per thousand. While growing to bait size these young pit-run worms should throw enough egg capsules to build up the breeder's own stock, a welcome "extra dividend".

This system of buying young bulk worms is good practice with either Africans or red worms, one that enables a beginning breeder to launch an advertising campaign and build up his customer list right from the start, and to supply all demands profitably while his own volume is building up.

ONE FURTHER SUGGESTION: If you are now a red worm raiser, don't get over-enthusiastic about Africans to the extent of giving up your red worm business completely in their favor. Work them simultaneously, if you choose, and make all the profit you can on both. In the final analysis, the good dependable easy-to-raise red worm is still, and we believe will continue to be, the backbone of the earthworm industry.

Nevertheless, the African Nightcrawler has gained a lot of favor, and is becoming so well known to the bait trade that no red worm grower can well afford to ignore it completely. He just might find it advisable, at some time in the future, to invoke that famous old admonition: "If you can't lick 'em, JOIN 'em!"

A FINAL THOUGHT: If you have a yen to raise this big tropical worm, we suggest that you first experiment with it on a limited scale until you have learned at first hand about its habits, feeding and care, thus eliminating any danger of substantial losses. Or "job" Africans for a while, if you prefer, to get the feel of the market before you make any important investment.

INTERESTING FACTS ABOUT AFRICAN NIGHTCRAWLERS

1 - The African Nightcrawler is the ONLY large worm of which we know that breeds and multiplies with sufficient rapidity, in pits, to make it commercially profitable.

2 - Fully grown it may reach a length of 12 inches or more, but is at its best for both bait and breeding stock at a length of 6 or 7 inches; a length reached at four to six months from the capsule stage.

3 - It is prone to "crawl" (migrate) and must be restrained by barriers or by lights over the pits, particularly in dark or wet weather. Open pit breeders lose heavily from this tendency, but northern growers, with housed and uniformly heated pits, can control it more easily.

4 - Africans are grown for bait and breeding stock only, having no value for soil improvement except in their native tropical habitat.

5 - Africans may be raised anywhere, regardless of climate or temperatures, if proper conditions and heat controls are provided for them.

6 - The African is a voracious feeder. It requires several times the amount of feed consumed by the red worm, but gains weight and length rapidly in proportion to the feed used.

7 - The African is a good shipper, under favorable temperature conditions, but cannot be shipped safely in temperatures much below 50 degrees.

8 - It costs more to raise Africans, as compared to red worms, but the premium prices they command make the cost and the trouble well worth while. The demand far exceeds the supply at this time.

9 - Africans should never be refrigerated in storage; they should be held at temperatures above 50 degrees.

10 - Many dependable sources for Africans, for either bait or breeding stock, are listed in the Earthworm Buyer's Guide. (See the book section.)

Earthworm Books for Extra Profits

ONE of the finest "springboards" to volume earthworm sales, in the experience of many successful breeders, is the advance sale of earthworm books and manuals to customers and prospective customers . . . books through which any breeder or potential breeder may obtain, in a few hours of comfortable reading, all of the hard-won knowledge that pioneer earthworm breeders have wrested from many decades of costly "trial and error" methods, or from a hundred years of tedious scientific experiment.

All of the earthworm books described in these pages are available for resale, through established growers, at liberal wholesale prices which enable the distributor to make a welcome extra profit on his book sales . . . enough, in some instances, to pay his entire advertising expense. At the same time, such book sales do much to promote volume earthworm sales.

The books may be purchased for stock at the maximum discount for reshipment by the grower-distributor, or at just a slightly smaller discount (the cost of postage and envelope) for drop-shipment by the publisher. Complete details of this plan, together with samples and prices of book folders, may be had by writing to Shields Publications, P.O. Box 669, Eagle River, WI 54521.

One of the most satisfactory things about book sales is the fact that the sale of one book usually leads to repeat sales of one or more additional books . . . sometimes the entire "Earthworm Library", amounting to $50.00 or more. Most of the books you'll sell carry the complete book listings in the back. Average book sales, according to the records of the distributors, run around $15.00 to $30.00.

Attractive 6-page book folders for enclosure with your mailings, (where as a rule they "ride free" so far as postage is concerned), are available to you at cost.

BOOK POSTAGE RATES: Books have a special rate. It is cheaper than the regular parcel post rate. This applies mostly to those instances when you mail several books. The package is marked "Special 4th Class Rate - Books" and mailed at the book rate. This is a flat rate per pound to any city in the United States.

EARTHWORM BOOKS FOR EXTRA PROFITS

OVER $100,000 WORTH OF THESE BOOKS PER YEAR!

If you have any doubts about the present interest and the future possibilities in the earthworm industry, THIS IS YOUR ANSWER! Each year book sales grow larger, and more important to the distributors who offer them. You will find more detailed descriptions of the various titles on the following pages.

HERE ARE THE BEST OF THE EARTHWORM BOOKS

EARTHWORM BUYER'S GUIDE & DIRECTORY
Shields Publications. The only directory published for the earthworm industry . . . a valuable guide for wholesale and retail earthworm buyers, bait dealers, fishermen, buyers of breeding stock. It lists many earthworm hatcheries in the United States and Canada. It also carries the display advertising of many leading growers.

THE NIGHTCRAWLER MANUAL
by Ray Edwards. This newly revised edition explains in detail the nightcrawler story. The author tells how they live, how to harvest them and how to keep them in top condition until they are sold. This comprehensive manual of 128 pages contains more than 30 photographs and illustrations. It is a must for anyone who collects nightcrawlers for a profit or his own use.

EARTHWORM FEEDS & FEEDING
by Charlie Morgan. An outstanding research report based on many years work and experience solving worm breeding problems. Tells how to prevent mites, acid beds, protein poisoning. Answers questions about best feeds, the use of antibiotics. Describes method for making synthetic manure plus many other ideas and methods by leading worm growers. Its 84 pages are filled with vital information, a valuable reference for every worm grower.

EARTHWORM SELLING AND SHIPPING GUIDE
by Charlie Morgan. The definitive guide to disposing of your product at a profit. How to sell at home or across the nation. Whole chapters are devoted to: How To Write A Classified Ad That Sells, A Sales Letter That Sells, Licenses and Postal Regulations, Packages and Delivery Methods, Packaging Materials and Preparation. Also, where to get and how to use additional FREE information; where, how, and when to advertise; where to get packages, packaging materials and exactly how to use them. HOW TO SELL MORE WORMS!

HARNESSING THE EARTHWORM
by Thomas Barrett. Dr. Barrett describes his book as "A practical inquiry into soil building, soil conditioning, and plant nutrition through the action of earthworms, with instruction for the intensive propagation and use of domesticated earthworms in biological soil building." It has facts, figures and illustrations showing how earthworms can be applied to condition your soil for better plants, more beautiful gardens and richer crops.

HOW TO RAISE, STORE & SELL NIGHTCRAWLERS

by Charlie Morgan. This is about the tough, big northern nightcrawler so popular with fishermen in many areas, the worm that brings a premium price. Morgan tells how to raise them inside and outside, what beddings and feeds to use, how to harvest and store them, tips on selling, packaging and shipping. Its 40 pages are full of useful information every worm grower can use.

PROFITABLE EARTHWORM FARMING

by Charlie Morgan. A comprehensive treatise by a research biologist who is also a practical earthworm breeder. A detailed study of earthworm biology, history, types, classifications and reproductive processes . . . PLUS many usable suggestions for bedding, feeding, handling, housing, heating, grading and counting, packaging and selling. Chapters, also, on native and African Nightcrawlers, Mealworms, Crickets, Grasshoppers and live baits generally. It's one of the best!

RAISING FISHWORMS WITH RABBITS

by Howard "Lucky" Mays. The author, a breeder of champion rabbits, shares his experiences in establishing a profitable business of raising fishworms under rabbit hutches, an ideal combination. 64 pages, 21 photos.

THE WORM FARM

by Charlie Morgan. A diary. The complete story of establishing a successful worm farm and business. In these pages you will relive the experiences of this worm grower . . . the headaches and triumphs that were part of starting and building a $20,000 business. A "must" for anyone in the worm business. 76 pages, illustrated.

THE ABC's OF THE EARTHWORM BUSINESS

by Ruth Meyers. The author's second book is aptly titled, as it spells out the fundamentals of raising earthworms to make money. She takes the beginner through all phases of the worm business, from selecting breeding stock to the fine points of marketing, including the "little" things that can be so important to a beginner.

RAISING THE AFRICAN NIGHTCRAWLER

by Charlie Morgan. The only complete book yet written concerning this interesting and highly profitable BIG bait worm to which so many established growers are now turning for extra profits; a thoroughly detailed book of primary importance to anyone who plans to grow Africans, or experiment with them as an adjunct to his red worm business.

LET AN EARTHWORM BE YOUR GARBAGE MAN

This report ***by Home, Farm and Garden Research, Inc.*** is far more comprehensive than the title indicates. It includes an extensive treatise, generously illustrated, on the role of the earthworm in the soil, by Henry Hopp of the U.S. Department of Agriculture, an eminent authority on earthworms.

LARGER RED WORMS

by George H. Holwager. George Holwager's eminent success has been largely due to his ability to produce a LARGER red worm, FASTER, and at amazingly LOW COST. He now passes on to other growers the secrets of his revolutionary feeding and fattening methods . . . 36 pages, illustrated, of refreshingly different and completely dependable earthworm raising information.

A-WORMING WE DID GO

by Ruth Myers. An inspiring account of how a handicapped woman starting "on a shoestring" built one thousand worms into one of the largest and best-known worm farms. Her friendly, humorous style makes enjoyable reading as she instructs how to build bins, mix feeds, tackle enemies of the earthworm and gives invaluable tips on packaging, shipping, advertising and working with bait dealers. Photos show farm equipment and operation.

DEAR CHARLIE

Compiled by Patrick H. Shields. The world's most renowned wormologist responds to your questions about worm farming in this new 64-page book, a collection of correspondence between the late Charlie Morgan and worm farmers around the world. They ask a wide variety of questions and Charlie answers them in his inimitable style.

To order any of the earthworm books listed here, a current price list may be obtained from the distributor through whom you obtained this book, or from the publisher.

NEW BOOKS & VIDEO

RECYCLE WITH EARTHWORMS:
The Red Wiggler Connection

By Shelley Grossman & Toby Weitzel $10.00
This book is the best we've seen on the subject! Vermicomposting and raising earthworms has never been so easy to understand and employ. These two "master composters" share their wealth of knowledge with environment-conscious organic gardeners, hobbyists and commercial worm farmers. Novice or expert, you'll appreciate the precise, clear, personal instruction as the authors reveal their trade secrets. Tons of illustrations and an important trouble-shooting chapter, too! 100 pages.

THE RED WIGGLER CONNECTION - *Video*

By Shelley Grossman & Toby Weitzel $19.95
What a great companion to the book, *Recycle with Earthworms*! This shows you, without gimmicks or sales pitches, how to be successful in your vermicomposting and worm raising venture. A 31-minute, high quality video which the authors use to demonstrate the techniques and principles in their highly acclaimed work.

AS THE WORM TURNS:
New & Easy Methods for Raising Earthworms

By Roy & Dianne Fewell $8.00
This new 1998 edition with a cute name has some seriously unique and valuable information! The authors and their family have been raising big, beautiful, bait-sized red worms for nearly fifty years. If you are interested in raising worms for the much-in-demand bait market, or just want to maintain a hobby-sized farm, this book is packed with wonderful worm wisdom including a "Worm Resources" section with details on how to obtain some of those hard-to-find supplies for the earthworm industry. 64 pages.

The above titles are available
from the distributor through whom you obtained this book
or from the publisher.

Include $3.00 Postage and Handling for Each Order

INDEX